# Basic Laboratory Experiments for General, Organic, and Biochemistry

**SECOND EDITION**

**Joseph M. Landesberg**
Adelphi University

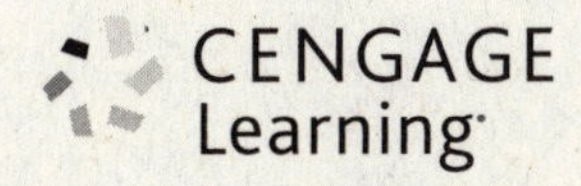

Australia • Brazil • Mexico • Singapore • United Kingdom • United States

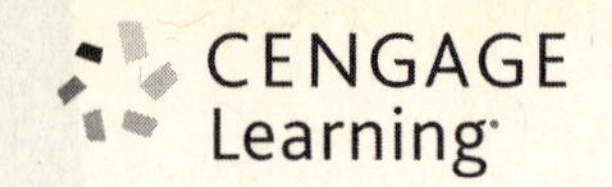

*Basic Laboratory Experiments for General, Organic, and Biochemistry*, 2e

WCN: 01-100-101

**Joseph M. Landesberg**

Product Manager: Mary Finch
Production Manager: David Woodbury
Content Developer: Elizabeth Woods
Content Coordinator: Elizabeth Woods
Product Assistant: Karolina Kiwak
Manufacturing Planner: Judy Inouye
Text and Cover Designer: PreMedia Global
Cover Image: © hiroyuki niu/iStock/Thinkstock



ISBN-13: 978-1-285-45965-3
ISBN-10: 1-285-45965-2

**Cengage Learning**
200 First Stamford Place, 4th Floor
Stamford, CT 06902
USA

Cengage Learning is a leading provider of customize solutions with office locations around the globe, inclu Singapore, the United Kingdom, Australia, Mexico, B Japan. Locate your local office at: **www.cengage.co**

Cengage Learning products are represented in Cana by Nelson Education, Ltd.

To learn more about Cengage Learning Solutions, vis **www.cengage.com**. Purchase any of our products a local college store or at our preferred online store **www.cengagebrain.com**.

Printed in the United States of America
1 2 3 4 5 6 7 18 17 16 15 14

# Contents

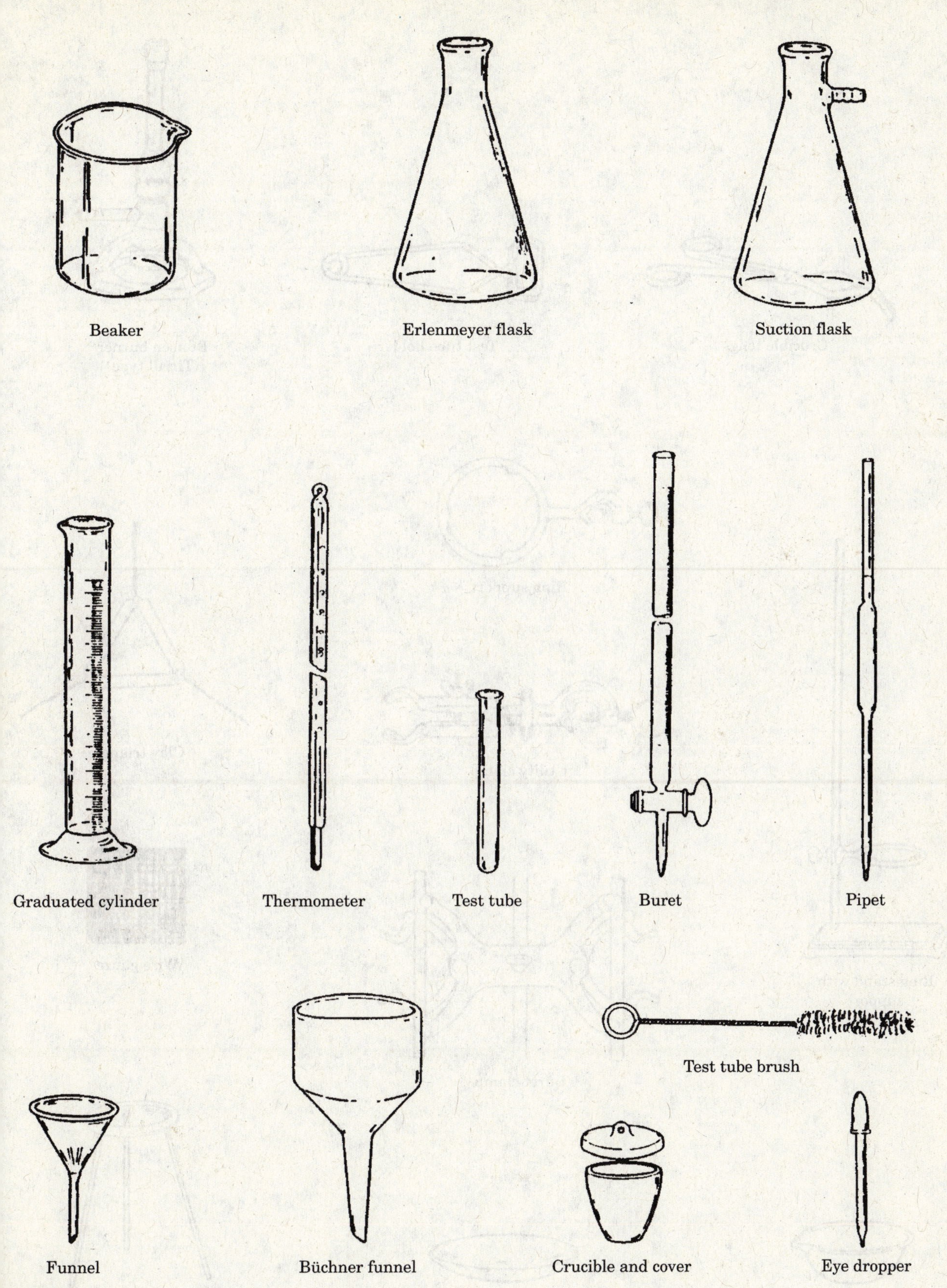


**Figure 1**
*Common laboratory equipment.*

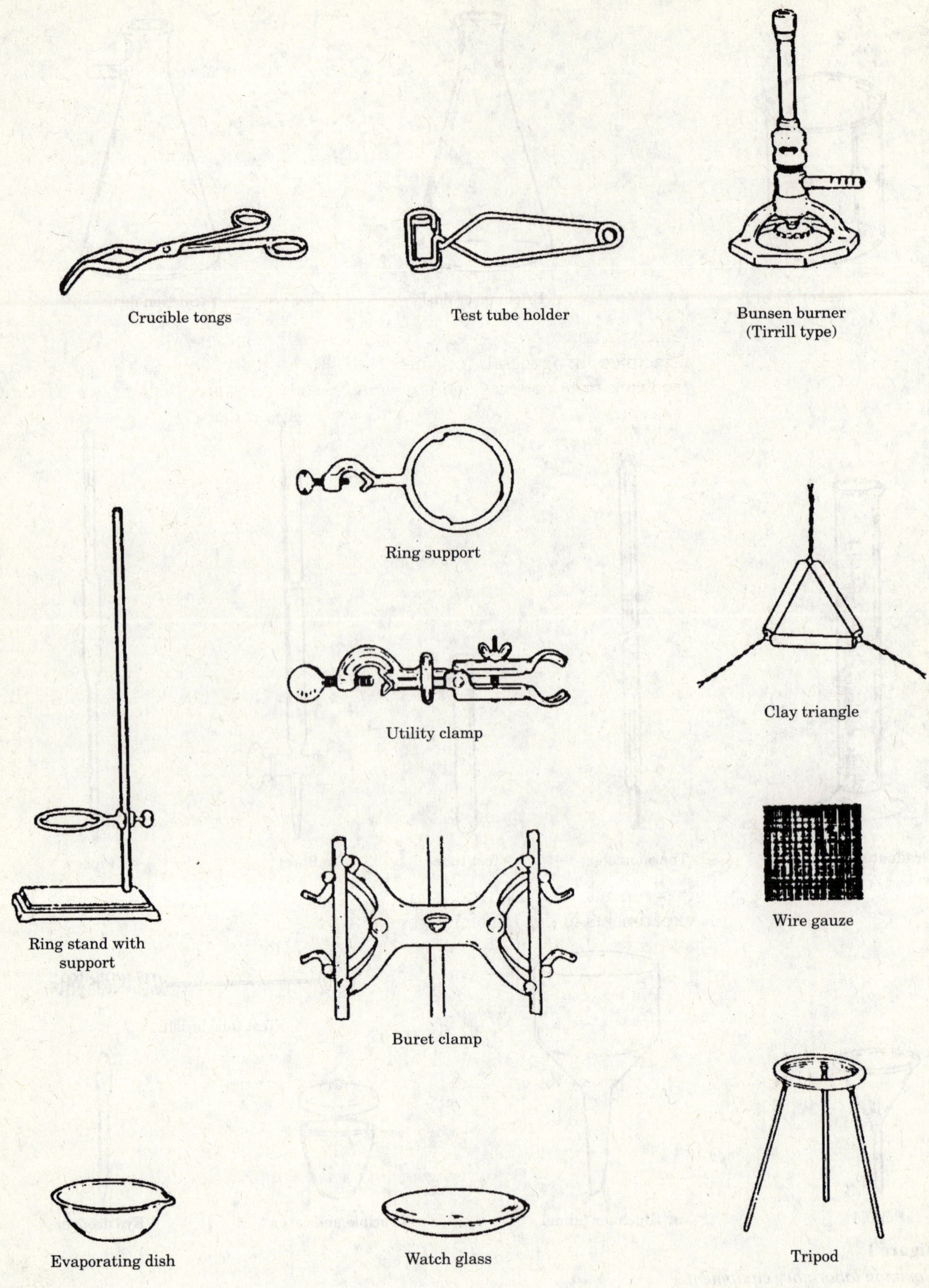
Crucible tongs
Test tube holder
Bunsen burner
(Tirrill type)
Ring support
Clay triangle
Utility clamp
Ring stand with
support
Wire gauze
Buret clamp
Evaporating dish
Watch glass
Tripod

**Figure 1**
*Continued*

# Preface

In preparing this Laboratory Manual, we used the experiments in *Laboratory Experiments for Introduction to General, Organic, and Biochemistry 8th Ed. (Frederick A. Bettelheim and Joseph M. Landesberg)* as the basis for the ones contained in this manual. This manual shares the outline and pedagogical philosophy with the larger edition. We have strived for the clearest possible writing in the procedures. The experiments give the student a meaningful, reliable laboratory experience that consistently works, while covering the basic principles of general, organic and biochemistry. Throughout the years, feedback from colleges and universities made us aware that we have managed to achieve manuals that not only ease the student's task in performing experiments, but also is student friendly. This revised manual maintains this standard.

This manual has improved procedures as a result of our observations of how our students carry out these experiments in our laboratories at Adelphi. We emphasize safety issues and waste disposal throughout. We strive to minimize the use of hazardous chemicals where possible and to design experiments that work either on a small scale or on a semi-micro scale. The Pre-Lab Questions prepare students for the day's exercises and the Post-Lab Questions review the lessons that the experiment teaches.

Three basic goals were followed in all the experiments: (1) the experiments should illustrate the basic concepts learned in the classroom; (2) the experiments should be clearly and concisely written so that students will easily understand the task at hand, will work with minimal supervision because the manual provides enough information on experimental procedures, and will be able to perform the experiments in a 2 ½-hr. laboratory period; (3) the experiments should not only be simple demonstrations, but also should contain a sense of discovery.

It did not escape our attention that in adopting this manual of laboratory experiments, the instructor must pay attention to budgetary constraints. All experiments in this manual generally use only inexpensive pieces of equipment and glassware, the most expensive may be a pH meter.

The 14 experiments in this book will provide the basic experiments for a one-semester course. The following are the principle features of this book:

1. The **Table of Contents** is organized so that the first 7 experiments illustrate the fundamentals of general chemistry and the next 7 those of organic and biochemistry.

2. Each experiment starts out with a **Background** which contains information beyond the textbook material. All the relevant principles and their applications for the experiment at hand are reviewed in this **Background** section.

3. The **Procedure** part provides a step-by-step description of the experiments. Clarity of writing in this section is of the utmost importance for successful

execution of the experiments. **Caution!** signs alert students when dealing with dangerous chemicals, such as strong acids or bases.

4. **Pre-Lab Questions** are provided to familiarize students with the concepts and procedures before they start the experiments. By requiring students to answer these questions and by grading their answers, we accomplish the task of preparing the students for the experiments.

5. In the **Report Sheet** we not only ask for the recording of raw data, but we also request some calculations to yield secondary data.

6. The **Post-Lab Questions** are designed so that students should be able to reflect upon the results, interpret them, and relate their significance.

7. At the end of the book in **Appendix 5**, we provide Stockroom Personnel with detailed instructions on preparation of solutions and other chemicals for each experiment. We also give detailed instructions as to how much material is needed for a class of 25 students.

An **Instructor's Manual** that accompanies this book is **solely for the use of the Instructor**. It helps in the grading process by providing ranges of the experimental results we obtained from class use. In addition it alerts the instructor to some of the difficulties that may be encountered in certain experiments.

The disposal of waste material is discussed for each experiment. For further information, we recommend *Prudent Practices in the Laboratory,* National Academy Press, Washington, DC (1995). A sample MSDS (Material Safety Data Sheet) is included to alert students to information regarding chemical safety. Laboratories should have these sheets on file for all chemicals that are used in the experiments, and these sheets should be made available to students on demand.

We hope that you will find our book of laboratory experiments helpful in instructing your students. We anticipate that students will find it inspiring in studying different aspects of chemistry.

Garden City, NY
September 2013 Joseph M. Landesberg

# Acknowledgments

These experiments have been used by our Adelphi colleagues over the years and their criticism and expertise were instrumental in the refinement of the experiments. We thank James Adamski, Frederick A. Bettelheim (dec.), Dennis Boyd, Stephen Freedman, Stephen Z. Goldberg, Robert W. Halliday (dec.), Cathy Ireland, Doug Kamen, Mahadevappa Kumbar (dec.), Robert Lippman, Jerry March (dec.), Sung Moon, Simeon Moshchitsky, Donald Opalecky, David Parkin, Michael Parris, Reuben Rudman (dec.), Charles Shopsis, Suzanne Sitkowski, Kevin Terrance, and Stanley Windwer for their suggestions. We thank Janet Maxwell, Angelo State University, for her helpful critique, also.

Of course, we could not be sure that the directions for solution preparation were clear were it not for the input of our stockroom staff, Mrs. Remany Abraham, Jonathan Roy, and Wesley Miller.

We extend our appreciation to the entire staff at Cengage Learning, especially to Ms. Elizabeth Woods, Associate Content Editor, for her encouragement and excellent efforts in producing this book.

# Practice Safe Laboratory

A few precautions can make the laboratory experience relatively hazard free and safe. These experiments are on a small-scale and as such, many of the dangers found in the chemistry laboratory have been minimized. In addition to specific regulations that you may have for your laboratory, the following **DO** and **DON'T RULES** should be observed at all times.

## DO RULES

- **Do wear approved safety glasses or goggles at all times.**
The first thing you should do after you enter the laboratory is to put on your safety eye-wear. The last thing you should do before you leave the laboratory is to remove them. Contact lens wearers must wear additional safety goggles; prescription glasses can be used instead. A chemical in the eye must be flushed with water for at least 15 minutes.

- **Do wear protective clothing.**
Wear sensible clothing in the laboratory: e.g., no shorts, no tank tops, no sandals. Be covered from the neck to the feet. Laboratory coats or aprons are recommended. Tie back long hair, out of the way of flames.

- **Do know the location and use of all safety equipment.**
This includes eyewash facilities, fire extinguishers, fire showers, and fire blankets. In case of fire, do not panic, clear out of the immediate area, and call your instructor for help. If clothes get on fire, the technique to extinguish flames is to **drop** and to **roll**.

- **Do use proper techniques and procedures.**
Closely follow the instructions given in this laboratory manual. These experiments have been student tested; however, accidents do occur but can be avoided if the steps for an experiment are followed. Pay heed to the **CAUTION!** signs in a procedure. Keep food and drinks out of the laboratory. Never taste, touch, or smell a chemical. Do not use your mouth to pipet a liquid.

- **Do discard waste material properly.**
Organic chemical waste should be collected in appropriate waste containers and <u>not flushed down sink drains</u>. Inorganic solutions, insoluble compounds, and toxic waste chemicals should be collected in properly labeled waste containers. Follow the directions of your instructor for alternate or special procedures. Discard glass in special glass containers.

- **Do be alert, serious, and responsible.**
Prepare for an experiment by carefully reading the procedure and being aware of the hazards before stepping foot into the laboratory. Keep your work area uncluttered.

## DON'T RULES

- **Do not eat or drink in the laboratory.**
  Consume any food or drink before entering the laboratory. Chemicals could get into food or drinks causing illness. If you must take a break, wash your hands thoroughly before leaving.

- **Do not smoke in the laboratory.**
  Smoke only in designated smoking areas outside the laboratory. Flammable gases and volatile flammable reagents could easily explode.

- **Do not taste any chemicals or breathe any vapors given off by a reaction.**
  If there is a need to smell a chemical, you will be shown how to do it safely.

- **Do not get any chemicals on your skin.**
  Wash off the exposed area with plenty of water should this happen. Notify your instructor at once. Wear gloves as indicated by your instructor.

- **Do not clutter your work area.**
  Your laboratory manual and the necessary chemicals, glassware, and hardware are all that should be on your benchtop. This will avoid spilling chemicals and breaking glassware.

- **Do not enter the chemical storage area or remove chemicals from the supply area.**
  Everyone must have access to the chemicals for the day's experiment. Removal of a chemical from the storage or supply area only complicates the proper execution of the experiment for the other students.

- **Do not perform unauthorized experiments.**
  Any experiment not authorized presents a hazard to any person in the immediate area.

- **Do not take unnecessary risks.**

These **DO** and **DON'T RULES** for a safe laboratory are not an exhaustive list, but are a minimum list of precautions that will make the laboratory a safe and fun activity. Should you have any questions about a hazard, ask your instructor first - not your laboratory partner. Finally, if you wish to know about the dangers of any chemical you work with, read the Material Safety Data Sheet (MSDS). These sheets should be on file in the chemistry department office.

SIGMA-ALDRICH

MATERIAL SAFETY DATA SHEET

Date Printed: 10/04/2004
Date Updated: 03/11/2004
Version 1.6

**Section 1 - Product and Company Information**

| | |
|---|---|
| Product Name | D- (+) -GLUCOSE |
| Product Number | G8270 |
| Brand | SIGMA |
| | |
| Company | Sigma-Aldrich |
| Street Address | 3050 Spruce Street |
| City, State, Zip, Country | SAINT LOUIS MO 63103 |
| US Technical Phone: | 314 771 5765 |
| Emergency Phone: | 414 273 3850 Ext. 5996 |
| Fax: | 800 325 5052 |

**Section 2 - Composition/Information on Ingredient**

| Substance Name | CAS # | SARA 313 |
|---|---|---|
| GLUCOSE | 50-99-7 | No |

Formula C6H12O6

Synonyms Anhydrous dextrose * Cartose * Cerelose * Corn sugar * Dextropur * Dextrose * Dextrose, anhydrous * Dextrosol * Glucolin * Glucose * Glucose, anhydrous * D-Glucose, anhydrous * Glucose liquid * Grape sugar * Sirup * Sugar, grape

RTECS Number: LZ6600000

**Section 3 - Hazards Identification**

HMIS RATING
HEALTH: 0
FLAMMABILITY: 0
REACTIVITY: 0

NFPA RATING
HEALTH: 0
FLAMMABILITY: 0
REACTIVITY: 0

For additional information on toxicity, please refer to Section 11.

**Section 4 - First Aid Measures**

ORAL EXPOSURE
If swallowed, wash out mouth with water provided person is conscious. Call a physician.

INHALATION EXPOSURE
If inhaled, remove to fresh air. If breathing becomes difficult, call a physician.

DERMAL EXPOSURE
In case of contact, immediately wash skin with soap and copious amounts of water.

EYE EXPOSURE
In case of contact with eyes, flush with copious amounts of water for at least 15 minutes. Assure adequate flushing by separating the eyelids with fingers. Call a physician.

## Section 5 - Fire Fighting Measures

FLASH POINT
N/A

AUTOIGNITION TEMP
N/A

FLAMMABILITY
N/A

EXTINGUISHING MEDIA
Suitable: Water spray. Carbon dioxide, dry chemical powder, or appropriate foam.

FIREFIGHTING
Protective Equipment: Wear self-contained breathing apparatus and protective clothing to prevent contact with skin and eyes. Specific Hazard(s): Emits toxic fumes under fire conditions.

## Section 6 - Accidental Release Measures

PROCEDURE(S) OF PERSONAL PRECAUTION(S)
Exercise appropriate precautions to minimize direct contact with skin or eyes and prevent inhalation of dust.

METHODS FOR CLEANING UP
Sweep up, place in a bag and hold for waste disposal. Avoid raising dust. Ventilate area and wash spill site after material pickup is complete.

## Section 7 - Handling and Storage

HANDLING
User Exposure: Avoid inhalation. Avoid contact with eyes, skin, and clothing. Avoid prolonged or repeated exposure.

STORAGE
Suitable: Keep tightly closed.

## Section 8 - Exposure Controls / PPE

ENGINEERING CONTROLS
Safety shower and eye bath. Mechanical exhaust required.

PERSONAL PROTECTIVE EQUIPMENT
Respiratory: Wear dust mask.
Hand: Protective gloves.
Eye: Chemical safety goggles.

GENERAL HYGIENE MEASURES
Wash thoroughly after handling.

## Section 9 - Physical/Chemical Properties

Appearance — Physical State: Solid

| Property | Value | At Temperature or Pressure |
|---|---|---|
| Molecular Weight | 180.16 AMU | |
| pH | N/A | |
| BP/BP Range | N/A | |
| MP/MP Range | 153 - 156 °C | |
| Freezing Point | N/A | |
| Vapor Pressure | N/A | |
| Vapor Density | N/A | |
| Saturated Vapor Conc. | N/A | |
| SG/Density | N/A | |
| Bulk Density | N/A | |
| Odor Threshold | N/A | |
| Volatile% | N/A | |
| VOC Content | N/A | |
| Water Content | N/A | |
| Solvent Content | N/A | |
| Evaporation Rate | N/A | |
| Viscosity | N/A | |
| Surface Tension | N/A | |
| Partition Coefficient | N/A | |
| Decomposition Temp. | N/A | |
| Flash Point | N/A | |
| Explosion Limits | N/A | |
| Flammability | N/A | |
| Autoignition Temp | N/A | |
| Refractive Index | N/A | |
| Optical Rotation | N/A | |
| Miscellaneous Data | N/A | |
| Solubility | N/A | |

N/A = not available

## Section 10 - Stability and Reactivity

STABILITY

Stable: Stable.
Materials to Avoid: Strong oxidizing agents.

HAZARDOUS DECOMPOSITION PRODUCTS

Hazardous Decomposition Products: Carbon monoxide, Carbon dioxide.

HAZARDOUS POLYMERIZATION

Hazardous Polymerization: Will not occur

## Section 11 - Toxicological Information

ROUTE OF EXPOSURE

Skin Contact: May cause skin irritation.
Skin Absorption: May be harmful if absorbed through the skin. Eye Contact: May cause eye irritation.
Inhalation: May be harmful if inhaled. Material may be irritating to mucous membranes and upper respiratory tract.
Ingestion: May be harmful if swallowed.

SIGNS AND SYMPTOMS OF EXPOSURE

To the best of our knowledge, the chemical, physical, and toxicological properties have not been thoroughly investigated.

TOXICITY DATA

Oral
Rat
25800 mg/kg
LD50
Remarks: Behavioral:Coma. Lungs, Thorax, or Respiration: Cyanosis. Gastrointestinal: Hypermotility, diarrhea.

Intraperitoneal
Mouse
18 GM/KG LD50

Intravenous
Mouse
9 GM/KG LD50

CHRONIC EXPOSURE - CARCINOGEN

Species: Rat
Route of Application: Subcutaneous
Dose: 15400 GM/KG Exposure Time: 22W Frequency: C
Result: Tumorigenic:Equivocal tumorigenic agent by RTECS criteria. Tumorigenic:Tumors at site or application.

CHRONIC EXPOSURE - TERATOGEN

Species: Woman
Dose: 2 GM/KG
Route of Application: Oral
Exposure Time: (28W PREG)
Result: Specific Developmental Abnormalities: Craniofacial (including nose and tongue). Specific Developmental Abnormalities: Other developmental abnormalities.

Species: Woman
Dose: 1057 UG/KG
Route of Application: Intravenous
Exposure Time: (39W PREG)
Result: Specific Developmental Abnormalities: Hepatobiliary system.

Species: Hamster
Dose: 20 GM/KG
Route of Application: Intraperitoneal
Exposure Time: (6-8D PREG)
Result: Specific Developmental Abnormalities: Eye, ear.

Species: Hamster
Dose: 20 GM/KG
Route of Application: Subcutaneous
Exposure Time: (6-8D PREG)
Result: Specific Developmental Abnormalities: Eye, ear.

Species: Hamster
Dose: 20 GM/KG
Route of Application: Multiple
Exposure Time: (6-8D PREG)
Result: Specific Developmental Abnormalities: Eye, ear. Specific Developmental Abnormalities: Musculoskeletal system.

CHRONIC EXPOSURE - MUTAGEN

Species: Human
Dose: 30 MMOL/L
Cell Type: Other cell types
Mutation test: DNA damage

Species: Mouse
Dose: 179 MMOL/L
Cell Type: lymphocyte
Mutation test: Mutation in mammalian somatic cells.

CHRONIC EXPOSURE - REPRODUCTIVE HAZARD

Species: Woman
Dose: 2 GM/KG
Route of Application: Intravenous
Exposure Time: (39W PREG)
Result: Maternal Effects: Other effects. Effects on Embryo or Fetus: Other effects to embryo.

Species: Woman
Dose: 1300 MG/KG
Route of Application: Intravenous
Exposure Time: (39W PREG)
Result: Effects on Newborn: Biochemical and metabolic. Effects on Newborn: Behavioral.

Species: Rat
Dose: 300 GM/KG
Route of Application: Intraperitoneal
Exposure Time: (30D PRE)
Result: Maternal Effects: Ovaries, fallopian tubes. Maternal Effects: Uterus, cervix, vagina.

Species: Hamster
Dose: 20 MG/KG
Route of Application: Multiple
Exposure Time: (6-8D PREG)
Result: Effects on Fertility: Post-implantation mortality (e.g., dead and/or resorbed implants per total number of implants). Specific Developmental Abnormalities: Urogenital system.

Section 12 - Ecological Information

No data available.

Section 13 - Disposal Considerations

APPROPRIATE METHOD OF DISPOSAL OF SUBSTANCE OR PREPARATION
Contact a licensed professional waste disposal service to dispose of this material. Dissolve or mix the material with a combustible solvent and burn in a chemical incinerator equipped with an afterburner and scrubber. Observe all federal, state, and local environmental regulations.

Section 14 - Transport Information

DOT
Proper Shipping Name: None
Non-Hazardous for Transport: This substance is considered to be non-hazardous for transport.

IATA
Non-Hazardous for Air Transport: Non-hazardous for air transport.

Section 15 - Regulatory Information

UNITED STATES REGULATORY INFORMATION
SARA LISTED: No
TSCA INVENTORY ITEM: Yes

CANADA REGULATORY INFORMATION
WHMIS Classification: This product has been classified in accordance with the hazard criteria of the CPR, and the MSDS contains all the information required by the CPR.
DSL: Yes
NDSL: No

Section 16 - Other Information

DISCLAIMER
For R&D use only. Not for drug, household or other uses.

WARRANTY
The above information is believed to be correct but does not purport to be all inclusive and shall be used only as a guide. The information in this document is based on the present state of our knowledge and is applicable to the product with regard to appropriate safety precautions. It does not represent any guarantee of the properties of the product. Sigma-Aldrich Inc., shall not be held liable for any damage resulting from handling or from contact with the above product. See reverse side of invoice or packing slip for additional terms and conditions of sale.
Copyright 2004 Sigma-Aldrich Co. License granted to make unlimited paper copies for internal use only.

# Laboratory Measurements

## OBJECTIVES

1. To learn how to use simple, common equipment found in the laboratory.
2. To learn to take measurements.
3. To be able to record these measurements with precision, with accuracy, and with the proper number of significant figures.
4. To apply the factor-label method for measurement conversions.

## BACKGROUND

At some point today, whether it is for this laboratory or for some other activity, you will probably be required to write down some information. That information might be in the form of a measurement. And then you might need to use that measurement in some calculation. How good that calculation is depends on a number of factors. This experiment will help you sort out those factors that make for "good data" and a "good result."

### Units of Measurement

The metric system of weights and measures is used by scientists of all fields, including chemists. This system uses the base 10 for the measurements; for conversions, measurements may be multiplied or divided by 10. Table 1.1 lists the most frequently used factors in the laboratory which are based on powers of 10.

**Table 1.1** *Frequently Used Factors*

| *Prefix* | *Power of 10* | *Decimal Equivalent* | *Abbreviation* |
|---|---|---|---|
| Micro | $10^{-6}$ | 0.000001 | μ |
| Milli | $10^{-3}$ | 0.001 | m |
| Centi | $10^{-2}$ | 0.01 | c |
| Kilo | $10^{3}$ | 1000 | k |

The measures of length, volume, mass, energy, and temperature are used to evaluate our physical and chemical environment. Table 1.2 compares the metric system with the more recently accepted SI system (International System of Units). The laboratory equipment associated with obtaining these measures is also included.

**Table 1.2** *Units and Equipment*

| *Measure* | *SI Unit* | *Metric Unit* | *Equipment* |
|---|---|---|---|
| Length | Meter (m) | Meter (m) | Meterstick |
| Volume | Cubic meter ($m^3$) | Liter | Pipet, graduated cylinder, Erlenmeyer flask, beaker |
| Mass | Kilogram (kg) | Gram (g) | Balance |
| Energy | Joule (J) | Calorie (cal) | Calorimeter |
| Temperature | Kelvin (K) | Degree Celsius (°C) | Thermometer |

***Accuracy, precision, and significant figures***

Chemistry is an empirical science, meaning it is a field of study that depends on experience and observation for data. An experiment that yields data requires the appropriate measuring devices in order to get accurate measurements. Once the data is in hand calculations are done with the numbers obtained. How good the calculations are depends on a number of factors: (1) how careful you are in taking the measurements (laboratory techniques); (2) how good your measuring device is in getting a true measure (accuracy); (3) how reproducible is the measurement (precision).

The measuring device usually contains a scale. The scale, with its subdivisions or graduations, tells the limits of the device's accuracy. You cannot expect to obtain a measurement better than your instrument is capable of reading. Consider the portion of the ruler shown in Fig. 1.1.

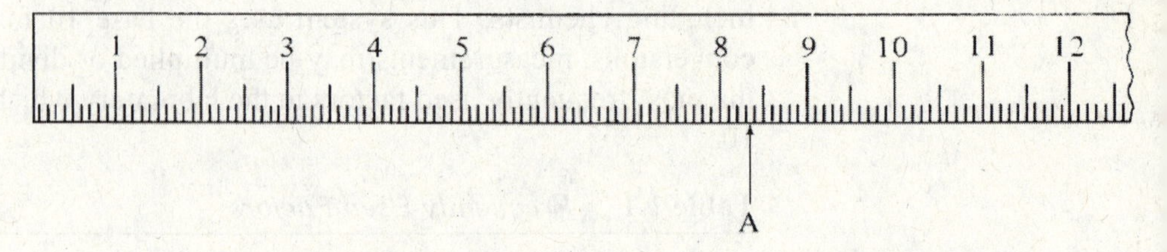


**Figure 1.1**
*Reading a metric ruler.*

There are major divisions labeled at intervals of 1 cm and subdivisions of 0.1 cm or 1 mm. The precision of the ruler is to 0.1 cm (or 1 mm); that is the measurement that is known for certain. However, it is possible to estimate to 0.01 cm (or 0.1 mm) by reading in between the subdivisions; this number is less accurate and of course, is less certain. In general, you should be able to record the measured value to one more place than the scale is marked. For example, in Fig. 1.1 there is a reading marked on the ruler. This value is 8.35 cm: two numbers are known with certainty, *8.3*, and one number, 0.0*5*, is uncertain since it is the *best estimate* of the fractional part of the subdivision. The number recorded, 8.35, contains 3 significant figures, 2 certain plus 1 uncertain. When dealing with *significant figures*, remember: (1) the uncertainty is in the last recorded digit, and (2) the number of significant figures contains the number of digits definitely known, plus one more that is estimated. The manipulation of significant figures in multiplication, division, addition, and subtraction is important. It is particularly important when using electronic calculators which give many more digits than are useful or significant. If you keep in mind the principle that the final answer can be no more accurate than the least accurate measurement, you should not go wrong. A few examples will demonstrate this.

*Example 1*

Divide 9.3 by 4.05. If this calculation is done by calculator, the answer found is 2.296296296. However, *a division should have as an answer the same number of significant figures as the least accurately known (fewest significant figures) of the numbers being divided.* One of the numbers, 9.3, contains only 2 significant figures. Therefore, the answer can only have 2 significant figures, i.e., 2.3 (rounded off).

*Example 2*

Multiply 0.31 by 2.563. Using a calculator, the answer is 0.79453. *As in division, a multiplication can have as an answer the same number of significant figures as the least accurately known (fewest significant figures) of the number being multiplied.* The number 0.31 has 2 significant figures (the zero only fixes the decimal point), therefore, the answer can only have 2 significant figures, i.e., 0.79 (rounded off).

*Example 3*

Add 3.56 + 4.321 + 5.9436. Calculation gives 13.8246. *With addition (or subtraction), the answer is significant to the least number of decimal places of the numbers added (or subtracted).* The least accurate number is 3.56, measured only to the hundredth's place. The answer should be to this accuracy, i.e., 13.82 (rounded off to the hundredth's place).

Finally, how do precision and accuracy compare? *Precision* is a determination of the reproducibility of a measurement. It tells you how closely several measurements agree with one another. Several measurements of the same quantity showing high precision will cluster together with little or no variation in value; however, if the measurements show a wide variation, the precision is low. *Random errors* are errors which lead to differences in successive values of a measurement and affect precision; some values will be off in one direction or another. One can estimate the precision for a set of values for a given quantity as follows: estimate = $\pm\Delta/2$, where $\Delta$ is the difference between the highest and lowest values.

*Accuracy* is a measure of how closely the value determined agrees with a known or accepted value. Accuracy is subject to *systematic errors*. These errors cause measurements to vary from the known value and will be off in the same direction, either too high or too low. A consistent error in a measuring device will affect the accuracy, but always in the same direction. It is important to use properly calibrated measuring devices. If a measuring device is not properly calibrated, it may give high precision, but with none of the measurements being accurate. However, a properly calibrated measuring device will be both precise and accurate. (See Fig. 1.2) A systematic error is expressed as the difference between the known value and the average values obtained by measurements in a number of trials.

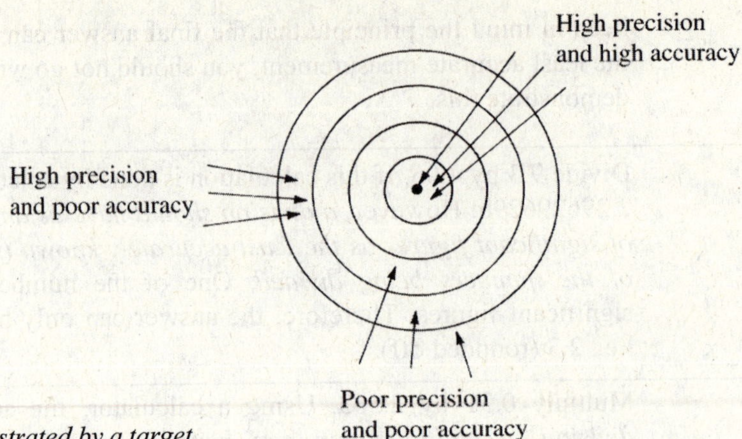


**Figure 1.2**
*Precision and accuracy illustrated by a target.*

## PROCEDURE

### Laboratory Measurements

***Length: use of the meterstick (or metric ruler)***

1. The meterstick is used to measure length. Examine the meterstick in your kit. You will notice that one side has its divisions in inches (in.) with subdivisions in sixteenths of an inch; the other side is in centimeters (cm) with subdivisions in millimeters (mm). Some useful conversion factors are listed below.

| | | | | | |
|---|---|---|---|---|---|
| 1 km | = | 1000 m | 1 in. | = | 2.54 cm |
| 1 m | = | 100 cm | 1 ft. | = | 30.48 cm |
| 1 cm | = | 10 mm | 1 yd. | = | 91.44 cm |
| 1 m | = | 1000 mm | 1 mi. | = | 1.61 km |

A meterstick that is calibrated to 0.1 cm can be read to the hundredth's place; however, only a 0 (0.00) or a 5 (0.05) may appear. A measurement falling directly on a subdivision is read as a 0 in the hundredth's place. A measurement falling anywhere between adjacent subdivisions is read as a 5 in the hundredth's place.

2. With your meterstick (or metric ruler), measure the length and width of this laboratory manual. Take the measurements in inches (to the nearest sixteenth of an inch) and in centimeters (to the nearest 0.05 cm). Record your response on the Report Sheet (1).
3. Convert the readings in cm to mm and m (2).
4. Calculate the area of the manual in $in^2$, $cm^2$, and $mm^2$ (3). Be sure to express your answers to the proper number of significant figures.

*Example 4* A student measured a piece of paper and found it to be 20.30 cm by 29.25 cm. The area was found to be 20.30 cm × 29.25 cm = 593.775 $cm^2$ = 593.8 $cm^2$ (rounded off to the proper number of significant figures).

***Volume: use of a graduated cylinder, an Erlenmeyer flask, and a beaker***

1. Volume in the metric system is expressed in liters (L) and milliliters (mL). Another way of expressing milliliters is in cubic centimeters ($cm^3$ or cc). Several conversion factors for volume measurements are listed below.

| | | | | | |
|---|---|---|---|---|---|
| 1 L | = | 1000 mL | 1 qt. | = | 0.95 L |
| 1 mL | = | 1 $cm^3$ = 1 cc | 1 gal. | = | 3.79 L |
| 1 L | = | 0.26 gal. | 1 fl. oz. | = | 29.6 mL |

2. The graduated cylinder is a piece of glassware used for measuring the volume of a liquid. Graduated cylinders come in various sizes with different degrees of accuracy. A convenient size for this experiment is the 100-mL graduated cylinder. Note that this cylinder is marked in units of 1 mL; major divisions are of 10 mL and subdivisions are of 1 mL. Estimates can be made to the nearest 0.1 mL. When a liquid is in the graduated cylinder, you will see that the level in the cylinder is curved with the lowest point at the center. This is the *meniscus,* or the dividing line between liquid and air. When reading the meniscus for the volume, be sure to read the *lowest* point on the curve and not the upper edge. To avoid errors in reading the meniscus, the eye's line of sight must be perpendicular to the scale (Fig. 1.3).

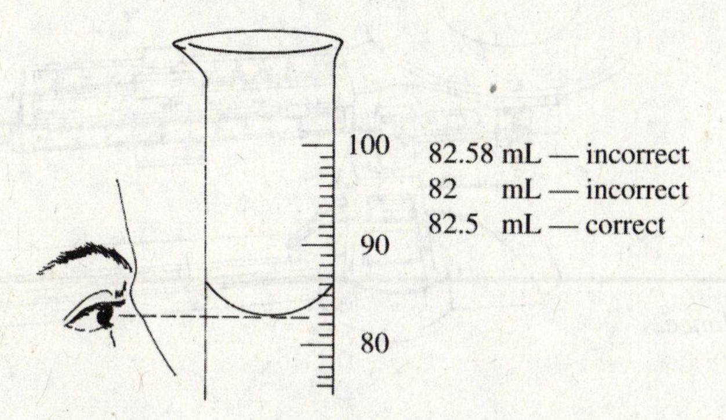


**Figure 1.3**
*Reading the meniscus on a graduated cylinder.*

3. Take a 50-mL graduated Erlenmeyer flask (Fig. 1.4) and fill with water to the 50 mL mark. Transfer the water, completely and without spilling, to a 100-mL graduated cylinder. Record the volume on the Report Sheet (4) to the nearest 0.1 mL; convert to L.

4. Take a 50-mL graduated beaker (Fig. 1.5), and fill with water to the 40 mL mark. Transfer the water, completely and without spilling, to a dry 100-mL graduated cylinder. Record the volume on the Report Sheet (5) to the nearest 0.1 mL; convert to L.

5. What is the error in mL and in percent for obtaining 50.0 mL for the Erlenmeyer flask and 40.0 mL for the beaker (6)? Calculate the % error in the following way: [(volume by graduated cylinder) minus (volume using the mark of the Erlenmeyer or the beaker) divided by (volume by graduated cylinder)] times 100.

6. Which piece of glassware will give you a more accurate measure of liquid: the graduated cylinder, the Erlenmeyer flask, or the beaker (7)?

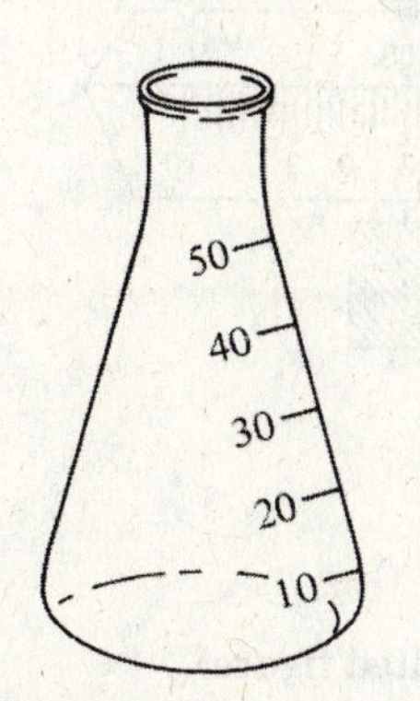


**Figure 1.4**
*A 50-mL graduated Erlenmeyer flask.*

### *Mass: use of the laboratory balance*

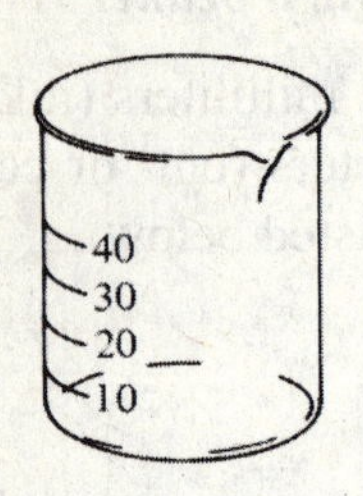


**Figure 1.5**
*A 50-mL graduated beaker.*

1. Mass measurements of objects are carried out with the laboratory balance. Many types of balances are available for the laboratory use. The proper choice of a balance depends upon what degree of precision you need for a measurement. The standard unit of mass is the kilogram (kg) in the SI system and the gram (g) in the metric system. Some conversion factors are listed below.

| | | | | | |
|---|---|---|---|---|---|
| 1 kg | = | 1000 g | 1 lb. | = | 454 g |
| 1 g | = | 1000 mg | 1 oz. | = | 28.35 g |

Three types of balances are illustrated in Fig. 1.6, 1.8, and 1.10. A platform triple beam balance is shown in Fig. 1.6. This balance can measure the mass of objects up to 2610 g. Since the scale is marked off in 0.1-g divisions, it is mostly used for a rough measure of mass; masses to 0.05 g can be estimated. Fig. 1.7 illustrates how to take a reading on this balance.

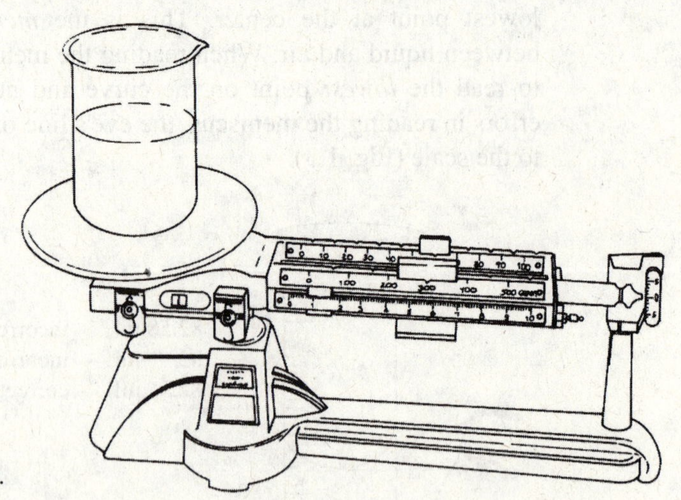

**Figure 1.6**
*A platform triple beam balance.*

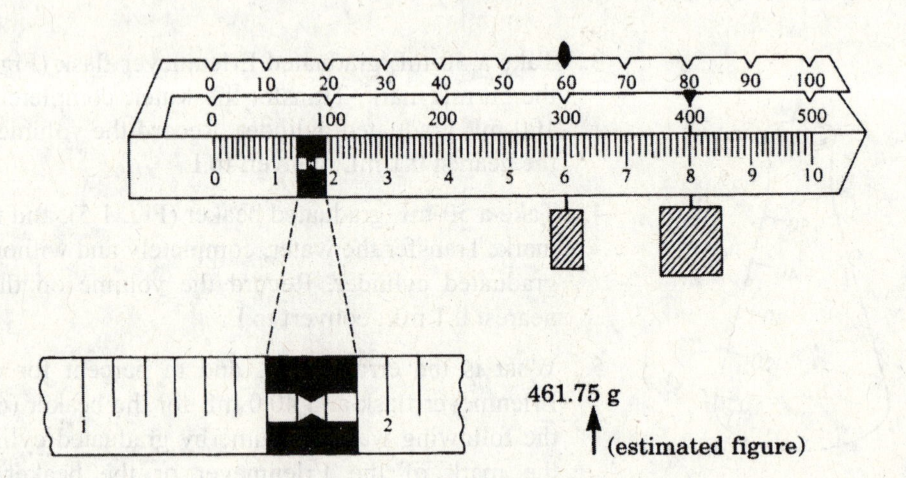


**Figure 1.7**
*Reading on a platform triple beam balance.*

The single pan, triple beam (Centigram®) balance is shown in Fig. 1.8. This Centogram® balance has a higher degree of accuracy since the divisions are marked in 0.01-g (estimates can be made to 0.005 g) increments.

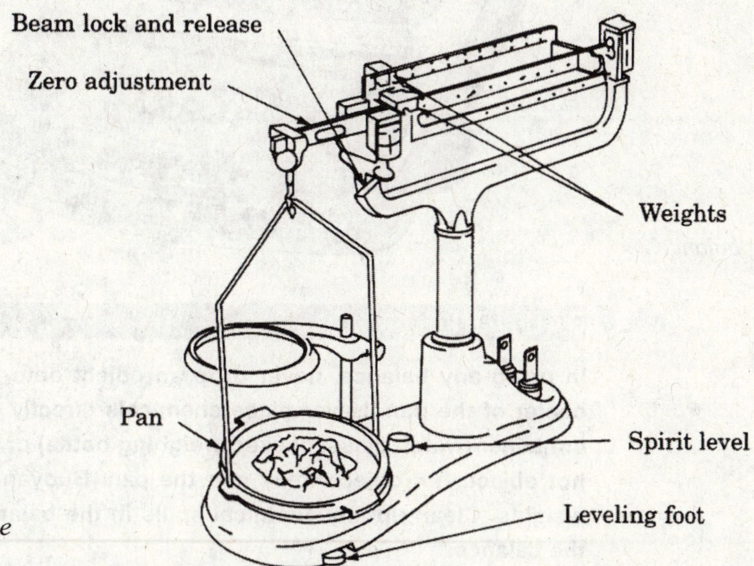


**Figure 1.8**
*A single pan, triple beam balance (Centogram®).*

Smaller quantities of material can be measured on this balance (to a maximum of 311 g). Fig. 1.9 illustrates how a reading on this balance would be read.

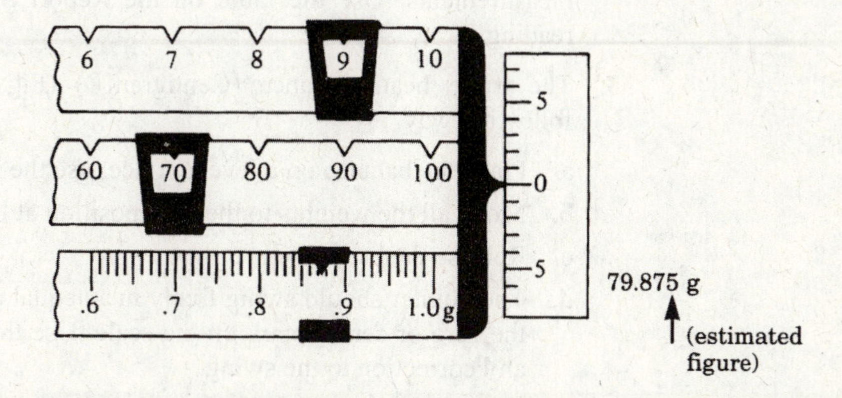


**Figure 1.9**
*Reading on a single pan, triple beam balance.*

Top-loading balances show the highest accuracy (Fig. 1.10). The mass of an object can be determined very rapidly with these balances because the total mass, to the nearest 0.001 g, can be read directly off the digital readout (Fig. 1.10). Balances of this type are very expensive and should be used only after the instructor has demonstrated their use.

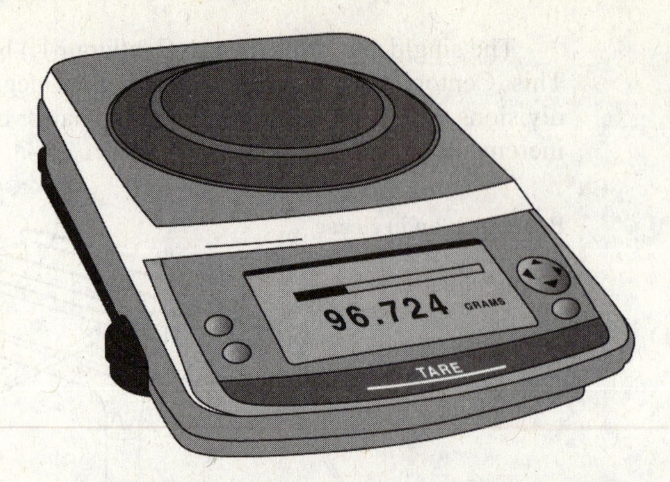


**Figure 1.10**
*A top-loading balance.*

**CAUTION**

**In using any balance, never drop an object onto the pan; place it gently in the center of the pan. Never place chemicals directly on the pan; use either a glass container (watch glass, beaker, weighing bottle) or weighing paper. Never weigh a hot object; hot objects may mar the pan. Buoyancy effects will cause incorrect weights. Clean up any chemical spills in the balance area to prevent damage to the balance.**

2. Measure the mass of a quarter, a test tube (13 × 100 mm) and a 125-mL Erlenmeyer flask. Express each mass reading to the proper number of significant figures. Use a platform triple beam balance, a single pan, triple beam balance (Centogram®), and a top-loading balance for these measurements. Use the table on the Report Sheet to record each mass reading.
3. The triple beam balance (Centigram®) (Fig. 1.8) is operated in the following way.
   a. Place the balance on a level surface; use the leveling foot to level.
   b. Move all the weights to the zero position at left.
   c. Release the beam lock.
   d. The pointer should swing freely in an equal distance up and down from the zero or center mark on the scale. Use the zero adjustment to make any correction to the swing.
   e. Place the object on the pan (remember the caution).
   f. Move the weight on the middle beam until the pointer drops; make sure the weight falls into the "V" notch. Move the weight back one notch until pointer swings up. This beam determines mass up to 10 g, in 1-g increments.
   g. Now move the weights on the back beam until the pointer drops; again be sure the weight falls into the "V" notch. Move the weight back one notch until the pointer swings up. This beam determines mass to 1 g, in 0.1-g increments.

h. Lastly, move the smallest weight (the cursor) on the front beam until the pointer balances, that is, swings up and down an equal distance from the zero or center mark on the scale. This last beam determines mass to 0.1 g, in 0.01-g increments.

i. The mass of the object on the pan is equal to the weights shown on each of the three beams (Fig.1.9). The mass to 0.005 g may be estimated.

j. Repeat the movement of the cursor to check your precision.

k. When finished, move the weights to the left, back to zero, and arrest the balance with the beam lock.

### *Temperature: use of the thermometer*

1. Routine measurements of temperature are done with a thermometer. Thermometers found in chemistry laboratories may use either mercury or a colored fluid as the liquid, and degrees Celsius (°C) as the units of measurement. The fixed reference points on this scale are the freezing point of water, 0 °C, and the boiling point of water, 100 °C. Between these two reference points, the scale is divided into 100 units, with each unit equal to 1 °C. Temperature can be estimated to 0.1 °C.
   Other thermometers use either the Fahrenheit (°F) or the Kelvin (K) temperature scale and use the same reference points, that is, the freezing and boiling points of water. Conversion between the scales can be accomplished using the formulas below.

$$°F = \frac{9}{5}°C + 32.0 \qquad °C = \frac{5}{9}(°F - 32.0) \qquad K = °C + 273.15$$

*Example 5* Convert 37.0 °C to °F and K

$$°F = \frac{9}{5}(37.0\ °C) + 32.0 = 98.6\ °F$$

$$K = 37.0\ °C + 273.15 = 310.2\ K$$

2. Use the thermometer in your kit and record to the nearest 0.1 °C the temperature of the laboratory at *room temperature*. Use the Report Sheet to record your results.

**CAUTION** 

**When you record temperature readings, be careful not to touch the walls of the beaker with the thermometer or hit the thermometer with the glass rod.**

3. Record the temperature of boiling water. Set up a 250-mL beaker containing 100 mL water, and heat on a hot plate until boiling. Hold the thermometer in the boiling water for at least 1 min. before reading the temperature. Using the Report Sheet, record your results to the nearest 0.1 °C

4. Record the temperature of ice water. Into a 250-mL beaker, add enough crushed ice to fill halfway. Add distilled water to the level of the ice. Stir

the ice water gently with a glass rod for 1 min. before placing the thermometer into the ice water. Hold the thermometer in the ice water for at least 1 min. before reading the temperature. Read the thermometer to the nearest 0.1 °C. Record your results on the Report Sheet.

**CAUTION** 

**When reading the thermometer, do not hold the thermometer by the mercury bulb. Body temperature will give an incorrect reading. If a mercury thermometer should break accidentally, call the instructor for proper disposal of the mercury. Mercury is toxic and very hazardous to your health. Do not handle the liquid or breathe its vapor.**

5. Convert your answers to questions 2, 3, and 4 into °F and K.

***Unit conversions and the factor-label method***

All of the scientific community, as well as most of the world, use the metric system for their measurements. Only a few, notably the United States and Great Britain, still use the English system as the primary system for weights and measure. Since we are caught is this dual environment that uses both of these systems, we frequently need to convert one measurement from one unit to another. Throughout the discussion above, there have been tables that contain equalities shown as *conversion factors*. The best way to use these factors is to employ the *factor-label method*. There is only one rule to follow when using this method: *when you multiply numbers, you also multiply the units; and when you divide numbers, you also divide the units.*

If you keep this rule in mind, the factor-label method lets you know when you have carried out a calculation correctly. Even though you may have entered all the digits correctly into your calculator, *if the units to the answer from the calculation are not the ones you are looking for, the calculation must be wrong.* If you follow the factor-label method correctly, you will get the correct mathematical solution. A few examples using this method will prepare you for problems in the **POST-LAB**.

*Example 6*

It is about 250 mi. from New York to Washington, D. C. How many kilometers is this?

The conversion factor (or equality) that applies to this problem is: 1 mi. = 1.61 km. There are 2 factors you can get from this:

$$\frac{1\text{ mi.}}{1.61\text{ km}} \text{ or } \frac{1.61\text{ km}}{1\text{ mi.}}.$$

So the solution is: $(250\text{ mi.}) \times \left(\frac{1.61\text{ km}}{1\text{ mi.}}\right) = 403\text{ km}$. Note that mi. cancels and you are left with the correct units.

If, on the other hand, you had used the other factor: $(250\text{ mi.}) \times \left(\frac{1\text{ mi.}}{1.61\text{ km}}\right) = 155\text{ mi}^2\text{/km}$, you get an answer that is incorrect since the units are incorrect (<u>even though the arithmetic is done correctly!</u>)

*Example 7*

Soda comes in a 1L bottle. How many quarts is this?

The equality: 1 qt. = 0.96 L; the factors: $\frac{1\text{ qt.}}{0.96\text{ L}}$ or $\frac{0.96\text{ L}}{1\text{ qt.}}$

Solution: $(1\text{ L}) \times \left(\frac{1\text{ qt.}}{0.96\text{ L}}\right) = 1.04\text{ qt.}$

*Example 8*

An Olympic athlete is entered in the 10,000 m race. How many miles is this race?

Equalities needed: 1 mi. = 1.6 km; 1 km = 1000 m.

Solution: $(10{,}000\text{ m}) \times \left(\frac{1\text{ km}}{1000\text{ m}}\right) \times \left(\frac{1\text{ mi.}}{1.6\text{ km}}\right) = 6.25\text{ mi.} = 6.3\text{ mi.}$

## CHEMICALS AND EQUIPMENT

1. 50-mL graduated beaker
2. 50-mL graduated Erlenmeyer flask
3. 100-mL graduated cylinder
4. Meterstick or ruler
5. Quarter
6. Balances
7. Hot plates
8. Test tube (13 × 100 mm)

name (print) ______ date (of lab meeting) ______ grade ______

course/section ______ partner's name (if applicable) ______

1 **EXPERIMENT 1: LABORATORY MEASUREMENTS**

# *Prelab Questions*

**A. Safety concerns.**

1. Why do you use a weighing container or weighing paper to hold a chemical when using a balance?

2. What precautions need to be followed when using a mercury thermometer?

**B. Basic principles.**

1. Why do scientists use the metric system of measurements instead of the English system?

2. Solve the following problems and record the answers to the proper number of significant figures:

   a. $50.2 \times 30.12 =$

   b. $9.03 \div 2.5 =$

   c. $5.03 + 6.059 + 1.003 =$

   d. $7.02 - 6.1 =$

3. Name the type of balance you would use for the following determinations:

   a. A mass of approximately 110 g ______

   b. An accurate mass of 110.000 g ______

name (print) ________ date (of lab meeting) ________ grade ________

course/section ________ partner's name (if applicable) ________

## 1 EXPERIMENT 1: LABORATORY MEASUREMENTS

# *Report Sheet*

**Length**

1. Length ________ in. ________ cm

   Width ________ in. ________ cm

2. Length ________ mm ________ m

   Width ________ mm ________ m

3. Area ________ $in^2$ ________ $cm^2$ ________ $mm^2$

   (Show calculations)

**Volume**

4. Erlenmeyer flask: volume in flask: 50 mL

   volume in graduated cylinder: ________ mL ________ L

5. Beaker: volume in beaker: 40 mL

   volume in graduated cylinder: ________ mL ________ L

6. Erlenmeyer flask

   Error in volume: volume in graduated cylinder – volume in flask = ________ mL

   $$\% \text{ Error} = \frac{\text{Error in volume}}{\text{Total volume}} \times 100 =$$ (Show your calculations) ________ %

   Beaker

   Error in volume: volume in graduated cylinder – volume in beaker = ________ mL

$$\% \text{ Error} = \frac{\text{Error in volume}}{\text{Total volume}} \times 100 =$$ (Show your calculations) ________ %

## Mass

| | **Balance** | | | | | |
|---|---|---|---|---|---|---|
| | *Platform* | | *Centogram®* | | *Top Loading* | |
| *Object* | *g* | *mg* | *g* | *mg* | *g* | *mg* |
| Quarter | | | | | | |
| Test Tube (13 × 100 mm) | | | | | | |
| 125-mL Erlenmeyer Flask | | | | | | |

## Temperature

| | *°C* | *°F* | *K* |
|---|---|---|---|
| Room Temperature | | | |
| Ice Water | | | |
| Boiling Water | | | |

How well do your thermometer readings agree with the accepted values for the freezing point and boiling point of water? Express any discrepancy as a deviation in degrees.

Deviation in Freezing Point (°C) ____________________

Deviation in Boiling Point (°C) ____________________

## Post-Lab Questions

1. From your results, which balance gave the most accurate measurement of mass?

2. A student attempted to find the mass of a warm beaker on a triple beam balance. What problems might the student encounter in trying to measure the mass of the beaker?

3. Two students each took measurements on a balance to determine the mass of a 125-mL Erlenmeyer flask that had a true mass of 70.621 g. Each student recorded three measurements for the flask and took an average. Below are the results.

| | Student A | Student B |
|---|---|---|
| | 70.519 | 70.596 |
| | 69.873 | 70.673 |
| | 70.934 | 70.643 |
| Average | 70.442 | 70.637 |

   a. Which set of results is more accurate?__________________________.

   b. Which set of results is more precise?___________________________.

   c. What can be said of the results from Student A and Student B?

4. A 250 mg sample was placed in a beaker weighing 15.645 g. What is the combined mass of the beaker and sample in grams?

5. Using your value for the mass of a quarter, how many (to the nearest whole number) would it take to make up a mass of one pound? Show your work.

6. Temperatures in the Southwest often reach 110°F in the summer. What is this temperature in °C? Show your work.

7. Mount Everest in the Himalayan Range is the highest peak in the world at 8850 m. What is this in (1) km and (2) mi.? Show your work.

8. A trip from Boston to Washington, D. C. is 450 mi. What is the distance in km? Show your work.

9. Measurements for a newborn were 24 in. and 9.98 lbs. What are the baby's measurements in cm and in kg? Show your work.

10. A container of corn oil reads "2 gal." on the label. How many quarts and how many liters are there in the container? Show your work.

EXPERIMENT 2

# *Density Determination*

## OBJECTIVES

1. To determine the densities of regular- and irregular-shaped objects and use them as a means of identification.
2. To determine the density of water.
3. To determine the density of a liquid and use this as a means of identification.

## BACKGROUND

How do you identify something or someone? You most likely would use some physical characteristics. If it were a person, for example, you probably would note the eyes, the hair color, or the facial features; for example, a green-eyed, red-haired girl, with freckles. In a similar way, matter can be identified by using its physical characteristics or properties. A substance may have a color, odor, melting point, or boiling point that is unique to it. These properties do not depend on the quantity of the substance and are called intensive properties. Density also is an intensive property and may serve as a means for identification.

The density of a substance is the ratio of its mass per unit volume. A value for density can be found by dividing the mass of a substance by its volume. The formula to remember is $\mathbf{d} = \frac{\mathbf{m}}{\mathbf{V}}$, where **d** is density, **m** is the mass, and **V** is the volume. While mass and volume do depend on the quantity of a substance (these are what we call extensive properties), the ratio is a constant at a given temperature. The units of density, reported in standard references, is in terms of g/mL (or g/cc or g/cm$^3$) at 20°C. The temperature is reported since the volume of a sample will change with temperature and, thus, so does the density.

*Example*

Divers recovered yellow bars from a sunken 16th century Spanish galleon. One of the bars had a mass of 453.6 g and a volume of 23.5 cm$^3$. Is the bar gold? (Density of gold = 19.3 g/cm$^3$ at 20°C.)

$$\mathbf{d} = \frac{\mathbf{m}}{\mathbf{V}} = \frac{\mathbf{453.6\ g}}{\mathbf{23.5\ cm^3}} = \mathbf{19.3\ g/cm^3}$$

Yes, it is gold.

## PROCEDURE

**Density of a Regular-Shaped Object**

1. Obtain a solid block from the instructor. Record the code number.
2. With your metric ruler, measure the dimensions of the block: length, width, height. Record the values to the nearest 0.05 cm (1, Trial 1).
3. Calculate the volume of the block: (length) × (width) × (height). Record the result (2, Trial 1).
4. Repeat the measurements and calculations for a Trial 2.
5. Using a balance (single pan, triple beam, or top-loading, whichever is available) determine the mass of the block. Record the mass to the nearest 0.05 g (3, Trial 1).
6. Calculate the density of the block (4, Trial 1).
7. Repeat the measurement and calculation for a Trial 2.

**Density of an Irregular-Shaped Object**

1. Obtain a sample of unknown substance from your instructor. Record the code number.
2. Obtain a mass of the sample of approximately 5 g. Be sure to record the exact quantity to the nearest 0.05 g (5, Trial 1).
3. Choose either a 10-mL or a 25-mL graduated cylinder. For small pieces, you should use the 10-mL graduated cylinder. For large pieces, you should use the 25-mL graduated cylinder. Fill the graduated cylinder approximately halfway with water. Record the volume to the nearest 0.05 mL (6, Trial 1). [Depending on the number of subdivisions between milliliter divisions, you can read to the hundredth's place. For example, you can read the 10-mL graduated cylinder to the hundredth's place where only a 0 (the reading is directly on the subdivision, e.g. 0.00) or a 5 (anywhere between two adjacent subdivisions, e.g. 0.05) can be used.]
4. Place the sample into the graduated cylinder. (If the pieces are too large for the opening of the 10-ml graduated cylinder, use the larger graduated cylinder.) Be sure all of the sample is below the water line. Gently tap the sides of the cylinder with your fingers to ensure that no air bubbles are trapped in the sample. Read the new level of the water in the graduated cylinder to the nearest 0.05 mL (7, Trial 1).
5. Calculate the volume of the sample (8, Trial 1) (Fig. 2.1). Assuming that the sample does not dissolve or react with the water, the difference between the two levels represents the volume of the sample.
6. Carefully recover the sample and dry it with a paper towel. Repeat the experiment for Trial 2.

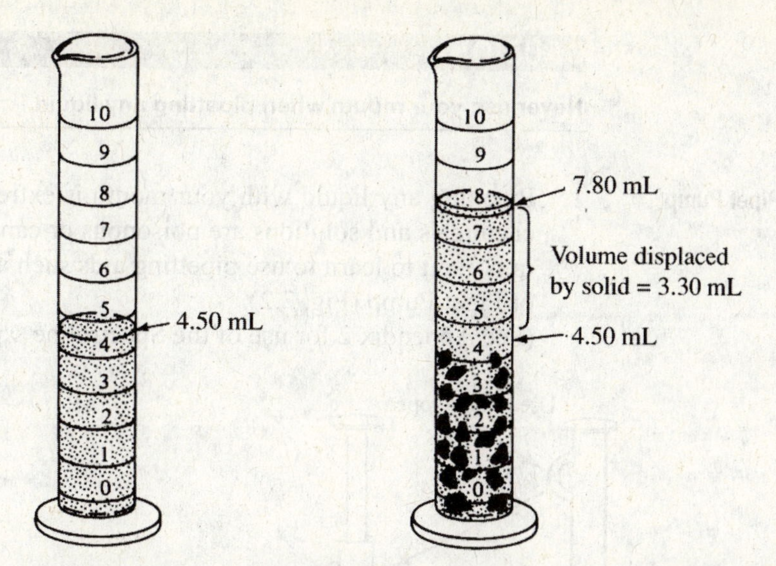


**Figure 2.1**
*Measurement of volume of an irregular-shaped object.*

7. Calculate the density of the sample from your data (9). From the results in trials 1 and 2, determine the average density. Report the result to the proper number of significant figures.
8. Determine the identity of your sample by comparing its density to the densities listed in Table 2.1 (10).
9. Recover your sample and return it as directed by your instructor.

**CAUTION!**

**Do not discard the samples in any general waste container or in the sink. Use a labeled collection container that is specific for each sample.**

**Table 2.1** *Densities of Selected Solids*

| *Sample* | *Density (g/cm$^3$)* |
|---|---|
| Wood (yellow pine) | 0.37 - 0.60 |
| Glass (common) | 2.5 - 2.8 |
| Aluminum (Al) | 2.70 |
| Zinc (Zn) | 7.13 |
| Tin (white) (Sn) | 7.29 |
| Iron (Fe) | 7.86 |
| Lead (Pb) | 11.30 |
| Gold (Au) | 19.31 |

**CAUTION!** 

**Never use your mouth when pipetting any liquid.**

**Use of the Pipet Pump**

1. Pipetting any liquid with your mouth is extremely dangerous. Many liquid chemicals and solutions are poisonous or can cause burns. That is why it is important to learn to use pipetting aids such as the Spectroline® pipet filler or Pipet Pump (Fig. 2.2).
   (See Appendix 2 for use of the Spectroline® pipet filler.)

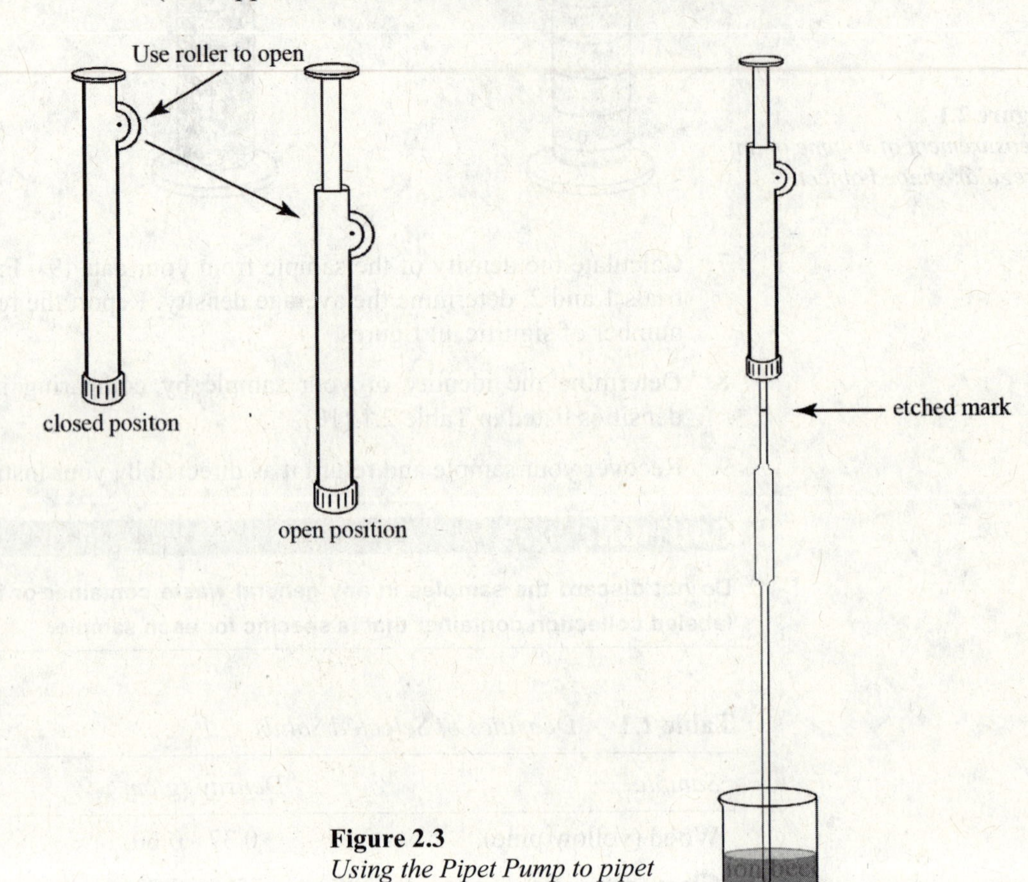


**Figure 2.2**
*The Pipet Pump.*

**Figure 2.3**
*Using the Pipet Pump to pipet liquid.*

2. Carefully insert the pipet end into the Pipet Pump (Fig. 2.3). The end should insert easily and not be forced.

**CAUTION!** 

**Before inserting the pipet end into the Pipet Pump, lubricate the glass by rubbing the opening with a drop of water or glycerine.**

3. Use the roller on the pump to draw liquid into the pipet. Allow the liquid to fill the pipet to a level slightly above the etched mark on the stem. Adjust the curved meniscus of the liquid to the etched mark (Fig. 2.4).

4. Withdraw the pipet from the liquid.
5. Drain the liquid from the pipet into a flask or beaker by using the roller. Continue to use the roller until no more liquid is expelled from the pipet. Remove any drops on the tip by touching the tip of the pipet against the inside wall of the collection flask. Liquid should remain inside the tip; the pipet is calibrated with this liquid in the tip. Do not blow out the excess.

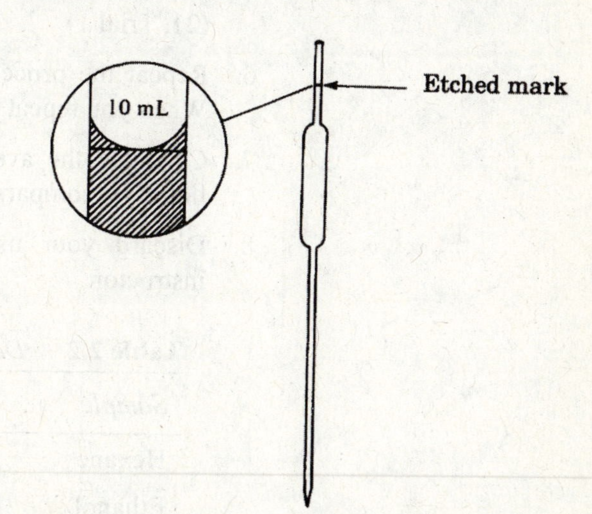


**Figure 2.4**
*The volumetric pipet; use this to measure water for the density determination of water*

**Density of Water**

1. Take a 100-mL beaker and fill it half-full with distilled water. Record the temperature of the water (11, Trial 1).
2. Determine the mass of a clean, dry 50-mL beaker to the nearest 0.001 g (12, Trial 1).
3. With a 5-mL volumetric pipet, transfer 5.00 mL of distilled water into the beaker from (1), above, using a Pipet Pump (Fig. 2.3). (Before transferring the distilled water, be sure there are no air bubbles trapped in the volumetric pipet. If there are, gently tap the pipet to dislodge the air bubbles, and then refill to the line.) Immediately determine the mass of the beaker and water; record the value to the nearest 0.05 g (13, Trial 1).
4. Calculate the mass of the water by subtraction (14, Trial 1).
5. Calculate the density of the water at the temperature recorded (15, Trial 1).
6. Repeat steps nos. 1, 2 , 3, 4, and 5 for Trial 2.
7. Calculate the average density (16). Compare your average value at the recorded temperature to the value reported for that temperature in a standard reference.

**Density of an Unknown Liquid**

1. Obtain approximately 25 mL of an unknown liquid from your instructor. Record the code number. Determine the temperature of the liquid (17, Trial 1).
2. Determine the mass of a clean, dry 50-mL beaker to the nearest 0.05 g (18, Trial 1).

3. Transfer 5.00 mL of the liquid with a 5-mL volumetric pipet into the beaker from (2), above, using the Pipet Pump (Fig. 2.3). Immediately determine the mass of the beaker and liquid to the nearest 0.05 g (19, Trial 1).
4. Calculate the mass of the unknown liquid by subtraction (20, Trial 1).
5. Calculate the density of the unknown liquid at the temperature recorded (21, Trial 1).
6. Repeat the procedure following steps nos. 1, 2, 3, 4, and 5 for Trial 2. When you repeat the steps, be sure all the glassware is clean and dry.
7. Calculate the average density. Determine the identity of your unknown liquid by comparing its density to the densities listed in Table 2.2 (22).
8. Discard your used liquid samples into containers provided by your instructor.

**Table 2.2** *Densities of Selected Liquids*

| *Sample* | *Density (g/mL)* | *T°C* |
|---|---|---|
| Hexane | 0.659 | 20 |
| Ethanol | 0.791 | 20 |
| Olive oil | 0.918 | 15 |
| Sea water | 1.025 | 20 (3.15% NaCl, g/100 g sol) |
| Milk | 1.028-1.035 | 20 |
| Ethylene glycol | 1.109 | 20 |

## CHEMICALS AND EQUIPMENT

1. Pipet Pump or Spectroline® pipet filler
2. 5-mL volumetric pipet
3. Solid wood block
4. Aluminum
5. Lead
6. Tin
7. Zinc
8. Hexane
9. Ethanol
10. Olive oil
11. Sea water
12. Milk
13. Ethylene glycol
14. Metric ruler
15. Balance (single pan, triple beam, or top loading)
16. 10-mL graduated cylinder
17. 25-mL graduated cylinder
18. 50-mL beaker

name (print) ______ date (of lab meeting) ______ grade ______

course/section ______ partner's name (if applicable) ______

2 **EXPERIMENT 2: DENSITY DETERMINATION**

# *Prelab Questions*

**A. Safety concerns.**

1. Why do you not discard your solid or liquid samples into the sink? Where should you discard these samples?

2. Should you use your mouth when you pipet a liquid? Explain.

3. Why is it necessary to lubricate the end of the pipet before inserting it into the pipet pump?

**B. Basic Principles.**

1. Identify the following characteristics as either an intensive property or an extensive property.

   a. Melting point______________________.

   b. Color______________________.

   c. Volume______________________.

   d. Mass______________________.

   e. Density______________________.

2. Cork stoppers float on water. Could you use the water displacement to determine the density of the cork stopper? Explain.

3. Exactly 50.0 mL of liquid has a mass of 40.30 g. What is its density? Show your work.

name (print) ______ date (of lab meeting) ______ grade ______

course/section ______ partner's name (if applicable) ______

**2** **EXPERIMENT 2: DENSITY DETERMINATION**

# *Report Sheet*

**Report all measurements and calculations to the correct number of significant figures.**

**A. Density of a regular-shaped object**

Unknown code number ______

| | Trial 1 | Trial 2 |
|---|---|---|
| 1. Length | ______ cm | ______ cm |
| Width | ______ cm | ______ cm |
| Height | ______ cm | ______ cm |
| 2. Volume (L × W × H) | ______ $cm^3$ | ______ $cm^3$ |
| 3. Mass | ______ g | ______ g |
| 4. Density: (3)/(2) | ______ g/cm | ______ $g/cm^3$ |

Average density of block ______ $g/cm^3$

**B. Density of an irregular-shaped object**

Unknown code number ______

| | Trial 1 | Trial 2 |
|---|---|---|
| 5. Mass of metal sample | ______ g | ______ g |
| 6. Initial volume of water | ______ mL | ______ mL |
| 7. Final volume of water | ______ mL | ______ mL |
| 8. Volume of metal: (7) – (6) | ______ mL | ______ mL |
| 9. Density of metal: (5)/(8) | ______ g/mL | ______ g/mL |

Average density of metal ______ g/mL

10. Identity of unknown metal ______

| **C. Density of water** | **Trial 1** | **Trial 2** |
|---|---|---|
| 11. Temperature of water | ________ °C | ________ °C |
| 12. Mass of 50-mL beaker | ________ g | ________ g |
| Volume of water | 5.00 mL | 5.00 mL |
| 13. Mass of beaker and water | ________ g | ________ g |
| 14. Mass of water: (13) – (12) | ________ g | ________ g |
| 15. Density of water: (14)/5.00 mL | ________ g/mL | ________ g/mL |

16. Average density of water ________ g/mL

Density found in literature ________ g/mL

| **D. Density of unknown liquid** | **Trial 1** | **Trial 2** |
|---|---|---|
| Unknown code number________ | | |
| 17. Temperature of unknown liquid | ________ °C | ________ °C |
| 18. Mass of 50-mL beaker | ________ g | ________ g |
| 19. Mass of beaker and liquid | ________ g | ________ g |
| 20. Mass of liquid: (19) – (18) | ________ g | ________ g |
| Volume of liquid | 5.00 mL | 5.00 mL |
| 21. Density of liquid: (20)/5.00 mL | ________ g/mL | ________ g/mL |

Average density of unknown liquid ________ g/mL

22. Identity of unknown liquid ________________

## Post-Lab Questions

1. When a student drew liquid into the volumetric pipet, air bubbles were trapped in the volumetric pipet. Would this give a density less than expected or greater than expected? Why?

2. A student has a regular wooden block to work with for a density determination. Unknown to the student is that the block has a hollow center. How will this affect the student's determination of the density?

3. Ethanol has a density of 0.791 g/cm$^3$ at 20 °C. How many milliliters (mL) are needed to have 30.0 g of liquid? Show your work.

## Post-Lab Questions

1. When a student was measuring the volume of a [illegible], bubbles were trapped in the [illegible]. Would the [illegible] density be [illegible] or greater than expected? Why?

2. A student has a regular wooden block to work with for the density experiment. Unknown to the student is that the block has a hollow center. How will this affect the student's determination of the density?

3. Ethanol has a density of 0.789 g/mL at 20°C. How many milliliters (mL) are needed to have 30.0 g of liquid? Show your work.

# Classes of Chemical Reactions

## OBJECTIVES

1. To demonstrate the different types of chemical reactions.
2. To be able to observe whether a chemical reaction has taken place.
3. To use chemical equations to describe a chemical reaction.

## BACKGROUND

The Periodic Table lists 114 named elements; an additional 4 synthetic elements have yet to be named. The chemical literature describes millions of compounds that are known- some isolated from natural sources, some synthesized by laboratory workers. The combination of chemicals, in the natural environment or the laboratory setting, involves chemical reactions. The change in the way that matter is composed is a *chemical reaction*, a process wherein reactants (or starting materials) are converted into products. The new products often have properties and characteristics that are entirely different from that of the starting materials.

Four ways in which chemical reactions may be classified are combination, decomposition, single replacement (substitution), or double replacement (metathesis).

Two elements reacting to form a compound is a *combination reaction*. This process may be described by the general formula:

$$A + B \rightarrow AB$$

The rusting of iron or the combination of iron and sulfur are good examples.

$$4Fe(s) + 3O_2(g) \rightarrow 2Fe_2O_3(s) \text{ (rust)}$$

$$Fe(s) + S(s) \rightarrow FeS(s)$$

Two compounds reacting together as in the example below also is a combination reaction.

$$CaO(s) + CO_2(g) \rightarrow CaCO_3(s)$$

A compound which breaks down into elements or simpler components typifies the *decomposition reaction*. This reaction has the general formula:

$$AB \rightarrow A + B$$

Some examples of this type of reaction are the electrolysis of water into hydrogen and oxygen:

$$2H_2O(l) \rightarrow 2H_2(g) + O_2(g)$$

and the decomposition of potassium iodate into potassium iodide and oxygen:

$$2KIO_3(s) \rightarrow 2KI(s) + 3O_2(g)$$

The replacement of one component in a compound by another describes the *single replacement* (or *substitution*) reaction. This reaction has the general formula:

$$AB + C \rightarrow CB + A$$

Processes which involve oxidation (the loss of electrons or the gain of relative positive charge) and reduction (the gain of electrons or the loss of relative positive charge) are typical of these reactions. Use of Table 3.1, the activity series of common metals, enables chemists to predict which oxidation-reduction reactions are possible. A more active metal, one higher in the table, is able to displace a less active metal, one listed lower in the table, from its aqueous salt. Thus, aluminum metal displaces copper metal from an aqueous solution of copper(II) chloride; but copper metal will not displace aluminum from an aqueous solution of aluminum(III) chloride.

$$2Al(s) + 3CuCl_2(aq) \rightarrow 3Cu(s) + 2AlCl_3(aq)$$

$$Cu(s) + AlCl_3(aq) \rightarrow \text{No Reaction}$$

(*Note that Al is oxidized to* $Al^{3+}$ *and* $Cu^{2+}$ *is reduced to Cu.*)

Hydrogen may be displaced from water by a very active metal. Alkali metals are particularly reactive with water, and the reaction of sodium with water often is exothermic enough to ignite the hydrogen gas released.

$$2Na(s) + 2HOH(l) \rightarrow 2NaOH(aq) + H_2(g) + \text{heat}$$

(*Note that Na is oxidized to* $Na^+$ *and* $H^+$ *is reduced to* $H_2$.)

Active metals, those above hydrogen in the series, are capable of displacing hydrogen from aqueous mineral acids such as HCl or $H_2SO_4$; but metals below hydrogen will not replace hydrogen. Thus zinc reacts with aqueous solutions of HCl and $H_2SO_4$ to release hydrogen gas, but copper will not.

$$Zn(s) + 2HCl(aq) \rightarrow ZnCl_2(aq) + H_2(g)$$

$$Cu(s) + H_2SO_4(aq) \rightarrow \text{No reaction}$$

**Table 3.1** *Activity Series of Common Metals*

| | | |
|---|---|---|
| K | (potassium) | Most active ↑ |
| Na | (sodium) | |
| Ca | (calcium) | |
| Mg | (magnesium) | |
| Al | (aluminum) | |
| Zn | (zinc) | |
| Fe | (iron) | Activity increases from bottom to top |
| Pb | (lead) | |
| $H_2$ | (hydrogen) | |
| Cu | (copper) | |
| Hg | (mercury) | |
| Ag | (silver) | |
| Au | (gold) | |
| Pt | (platinum) | Least active |

Two compounds reacting with each other to form two different compounds describes *double replacement* (or *metathesis*). This process has the general formula:

$$AB + CD \rightarrow AD + CB$$

There are two replacements in the sense that A replaces C in CD and C replaces A in AB. This type of reaction generally involves ions which form in solution either from the dissociation of ionic compounds or the ionization of molecular compounds. The reaction of an aqueous solution of silver nitrate with an aqueous solution of sodium chloride is a good example. The products are sodium nitrate and silver chloride. We know a reaction has taken place since the insoluble precipitate silver chloride forms and separates from solution.

$$AgNO_3(aq) + NaCl(aq) \rightarrow NaNO_3(aq) + AgCl(s)$$

(White precipitate)

In general, a double replacement results if one combination of ions leads to a precipitate, a gas or an un-ionized or very slightly ionized species such as water. In all of these reaction classes, it is very often possible to use your physical senses to observe whether a chemical reaction has occurred. The qualitative criteria may involve the formation of a gaseous product, the formation of a precipitate, a change in color, or a transfer of energy (the solution may become warmer or colder).

## PROCEDURE

**CAUTION!**

**Safety glasses must be worn at all times.**

**Combination Reactions**

1. Obtain a piece of aluminum foil approximately 2 × 0.5 in. Hold the foil at one end with a pair of forceps or crucible tongs and hold the other end in the hottest part of the flame of a Bunsen burner (see **Appendix 1**). Observe what happens to the foil. Has the metal changed in its appearance? Record your observation and complete a balanced equation for the reaction that has occurred (1). Place the foil on a wire gauze to cool.
2. Obtain a piece of copper foil approximately 2 × 0.5 in. (A copper penny, one minted before 1982, may be substituted.) Hold the foil at one end with a pair of forceps or crucible tongs and hold the other end in the hottest part of the flame of a Bunsen burner (see **Appendix 1**). Observe what happens to the metal. Has the metal changed in its appearance? Record your observation and complete a balanced equation for the reaction that has occurred (2). Place the foil on a wire gauze to cool.
3. Place the piece of aluminum foil from (1) into a test tube (13 × 100 mm). Add 1 mL of water, and with glass rod, crush the foil against the bottom of the test tube (be careful not to poke through the test tube bottom). Does any of the solid dissolve into the water? Record your observation (3).
4. Place the piece of copper foil from (2) into a test tube (13 × 100 mm). Add 1 mL of water, and with a glass rod, crush the foil against the bottom of the test tube (be careful not to poke through the test tube bottom). Does any of the solid dissolve into the water? Record your observation (4).

**Decomposition Reactions**

1. *Decomposition of ammonium carbonate.* Place 0.5 g of ammonium carbonate into a clean, dry test tube (13 × 100 mm). Gently heat the test tube in the flame of a Bunsen burner (Fig. 3.1). As you heat, hold a piece of wet red litmus paper with forceps just inside the mouth of the test tube. What happens to the solid? Are any gases produced? What happens to the color of the litmus paper? Ammonia gas acts as a base and turns moist red litmus paper blue. Record your observations and complete a balanced equation for the reaction that has occurred (5).

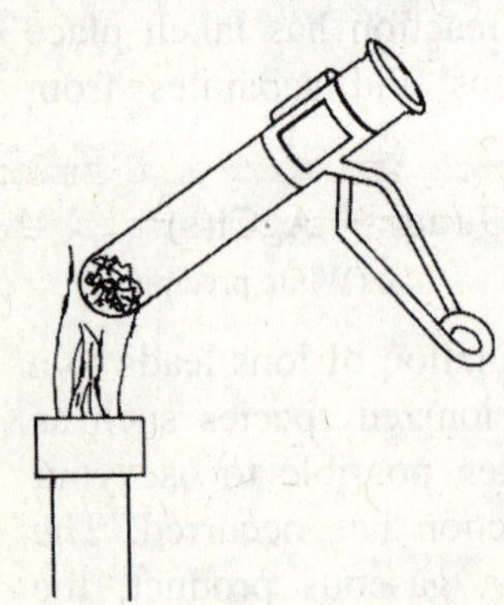

**Figure 3.1**
*Holding a test tube in a Bunsen burner flame.*

**CAUTION!**

**When heating the contents of a solid in a test tube, do not point the open end towards anyone.**

2. *Decomposition of potassium iodate.*
   a. Obtain three clean, dry test tubes (13 × 100 mm). Label them as no. 1, no. 2, and no. 3; add 0.5 g of compound according to the table below.

| *Test Tube No.* | *Compound* |
| --- | --- |
| 1 | $KIO_3$ |
| 2 | $KIO_3$ |
| 3 | KI |

b. With the test tube holder at the upper end of the test tube, heat test tube no. 1 with the hottest flame of the Bunsen burner as shown in Fig. 3.2. While test tube no. 1 is being heated, insert a glowing wooden splint into the opening of the test tube and move it half-way down. (The splint should not be flaming but should be glowing with sparks after the flame has been blown out. *Do not drop the glowing splint into the hot $KIO_3$.* Note the wooden splint is held by forceps.) Oxygen supports combustion. The glowing splint should glow brighter or may burst into flame in the presence of oxygen. Record what happens to the glowing splint (6). Remove the splint from the test tube.

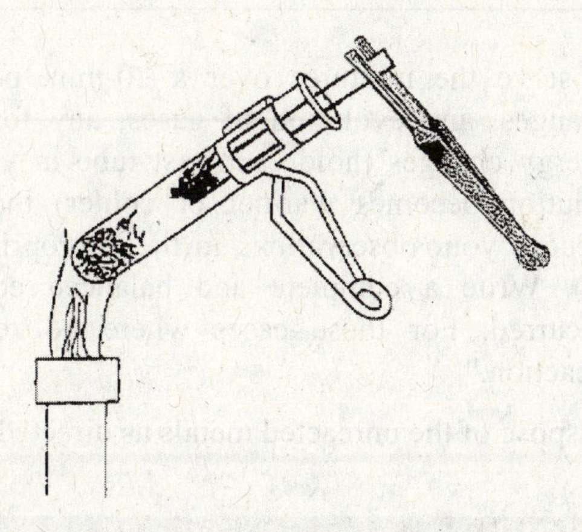

**Figure 3.2**
*Testing for oxygen gas.*

c. Remove the test tube from the flame and set it aside to cool.

d. Add 5 mL of distilled water to test tubes no. 1, no. 2, and no.3. Mix thoroughly to ensure that the solids are completely dissolved. Now add 10 drops of 0.1 M $AgNO_3$ solution to each test tube and shake. Observe what happens to each solution. Record the colors of the precipitates and write complete balanced equations for the reactions taking place in test tube no. 2 and test tube no. 3 (7). (You can tell the difference between the $KIO_3$ and KI solids by the test results with $AgNO_3$: $AgIO_3$ is a white precipitate; AgI is a yellow precipitate.)

e. Does a precipitate result when the residue in test tube no.1 is mixed with $AgNO_3$ (8)? What is the color of the precipitate? Based on the color of the precipitate, what compound is present in test tube no. 1 after heating $KIO_3$ (9)?

f. Write a complete balanced equation for the decomposition reaction (10).

**Single Replacement Reaction**

1. In a test tube rack, set up labeled test tubes (13 × 100 mm) numbered from 1 through 9. Place 1 mL (approx. 20 drops) of the appropriate solution in the test tube with a small piece of metal as outlined in the table below.

| *Test Tube No.* | *Solution* | *Metal* |
|---|---|---|
| 1 | $H_2O$ | Ca |
| 2 | $H_2O$ | Mg |
| 3 | $H_2O$ | Al |
| 4 | 3 M HCl | Zn |
| 5 | 6 M HCl | Pb |
| 6 | 6 M HCl | Cu |
| 7 | 0.1 M $NaNO_3$ | Al |
| 8 | 0.1 M $CuCl_2$ | Al |
| 9 | 0.1 M $AgNO_3$ | Cu |

2. Observe the mixtures over a 20-min. period of time. Note any color changes, any evolution of gases, any formation of precipitates, or any energy changes (hold each test tube in your hand and note whether the solution becomes warmer or colder) that occur during each reaction. Record your observations in the appropriate spaces on the Report Sheet (9). Write a complete and balanced equation for each reaction that occurred. For those cases where no reaction took place, write "No Reaction."

3. Dispose of the unreacted metals as directed by your instructor.

**CAUTION!**

**Do not discard any of the metals into the sink. Use specified waste containers.**

**Double Replacement Reaction**

1. Each experiment in this part requires mixing equal volumes of two solutions in a test tube (13 × 100 mm). Use about 10 drops of each solution. Record your observation at the time of mixing (10). When there appears to be no evidence of a reaction, feel the test tube for an energy change (exothermic or endothermic). The solutions to be mixed are outlined in the table on next page.

| *Test Tube No.* | *Solution No. 1* | *Solution No. 2* |
|---|---|---|
| 1 | 0.1 M $NaCl$ | 0.1 M $KNO_3$ |
| 2 | 0.1 M $NaCl$ | 0.1 M $AgNO_3$ |
| 3 | 0.1 M $Na_2CO_3$ | 3 M $HCl$ |
| 4 | 3 M $NaOH$ | 3 M $HCl$ |
| 5 | 0.1 M $BaCl_2$ | 3 M $H_2SO_4$ |
| 6 | 0.1 M $Pb(NO_3)_2$ | 0.1 M $K_2CrO_4$ |
| 7 | 0.1 M $Fe(NO_3)_3$ | 3 M $NaOH$ |
| 8 | 0.1 M $Cu(NO_3)_2$ | 3 M $NaOH$ |

2. For those cases where a reaction occurred, write a complete and balanced equation. Indicate precipitates, gases, and color changes. Table 3.2 lists some insoluble salts. For those cases where no reaction took place, write "No Reaction."
3. Discard the solutions as directed by your instructor.

**CAUTION!**

**Do not discard any of the solutions into the sink. Use specified waste containers.**

**Table 3.2** *Some Insoluble Salts*

| | |
|---|---|
| $AgCl$ | Silver chloride (white) |
| $Ag_2CrO_4$ | Silver chromate (red) |
| $AgIO_3$ | Silver iodate (white) |
| $AgI$ | Silver iodide (yellow) |
| $Ag_2SO_4$ | Silver sulfate (white) |
| $BaSO_4$ | Barium sulfate (white) |
| $Cu(OH)_2$ | Copper(II) hydroxide (blue) |
| $Fe(OH)_3$ | Iron(III) hydroxide (red) |
| $PbCrO_4$ | Lead(II) chromate (yellow) |
| $PbI_2$ | Lead(II) iodide (yellow) |
| $PbSO_4$ | Lead(II) sulfate (white) |

## CHEMICALS AND EQUIPMENT

1. Aluminum foil
2. Aluminum wire
3. Copper foil
4. Copper wire
5. Ammonium carbonate, $(NH_4)_2CO_3$
6. Potassium iodate, $KIO_3$
7. Potassium iodide, KI
8. Calcium turnings
9. Magnesium ribbon
10. Mossy zinc
11. Lead shot
12. 3 M HCl
13. 6 M HCl
14. 3 M $H_2SO_4$
15. 3 M NaOH
16. 0.1 M $AgNO_3$
17. 0.1 M NaCl
18. 0.1 M $NaNO_3$
19. 0.1 M $Na_2CO_3$
20. 0.1 M $KNO_3$
21. 0.1 M $K_2CrO_4$
22. 0.1 M $BaCl_2$
23. 0.1 M $Cu(NO_3)_2$
24. 0.1 M $CuCl_2$
25. 0.1 M $Pb(NO_3)_2$
26. 0.1 M $Fe(NO_3)_3$
27. Test tubes (13 × 100 mm)

name (print) ______ date (of lab meeting) ______ grade ______

course/section ______ partner's name (if applicable) ______

3 **EXPERIMENT 3: CLASSES OF CHEMICAL REACTIONS**

# *Prelab Questions*

**A. Safety concerns.**

1. Do you need to wear safety glasses for this experiment?

2. When heating a test tube, how do you position the open end?

**B. Basic principles.**

For each of the reactions below, classify as a combination, decomposition, single replacement or double replacement.

1. $Ba(s) + Cl_2(g) \rightarrow BaCl_2(s)$ ______

2. $2Ca(s) + O_2(g) \rightarrow 2CaO(s)$ ______

3. $CaCl_2(aq) + H_2SO_4(aq) \rightarrow 2HCl(aq) + CaSO_4(s)$ ______

4. $NH_3(aq) + HNO_3(aq) \rightarrow NH_4NO_3(aq)$ ______

5. $Hg(NO_3)_2(aq) + 2NaI(aq) \rightarrow HgI_2(s) + 2NaNO_3(aq)$ ______

6. $AgNO_3(aq) + KCl(aq) \rightarrow AgCl(s) + KNO_3(aq)$ ______

7. $Zn(s) + 2HCl(aq) \rightarrow ZnCl_2(aq) + H_2(g)$ ______

8. $H_2CO_3(aq) \rightarrow CO_2(g) + H_2O(l)$ ______

9. $2H_2O(l) \rightarrow 2H_2(g) + O_2(g)$ ______

10. $2Na(s) + 2H_2O(l) \rightarrow 2NaOH(aq) + H_2(g)$ ______

name (print) ______ date (of lab meeting) ______ grade ______

course/section ______ partner's name (if applicable) ______

## 3 EXPERIMENT 3: CLASSES OF CHEMICAL REACTIONS

# *Report Sheet*

Write *complete, balanced equations* for all cases that a reaction takes place. Your observation that a reaction occurred would be by a color change, by the formation of a gas, by the formation of a precipitate, or by an energy change (exothermic or endothermic). Those cases showing no evidence of a reaction, write "No Reaction."

## *Classes of Chemical Reactions*

**Combination reactions** **Observation**

1. _____ $Al(s)$ + _____ $O_2(g)$ → ______

2. _____ $Cu(s)$ + _____ $O_2(g)$ → ______

3. Solubility of aluminum oxide

4. Solubility of copper oxide ______

**Decomposition reactions**

5. _____ $(NH_4)_2CO_3(s)$ → ______

6. What happens to the glowing splint? ______

7. Observed colors in test tubes: no. 1 ______; no. 2 ______; no. 3 ______

_____ $KIO_3(aq)$ + _____ $AgNO_3(aq)$ → ______

_____ $KI(aq)$ + _____ $AgNO_3(aq)$ → ______

8. Results when residue of $KIO_3$ is mixed with $AgNO_3$ solution ______

9. The formula of the residue present after heating $KIO_3$. ______

10. _____ $KIO_3$ + heat → ______

**Single replacement reaction** **Observation**

11. *Test tube no.*

*1.* __Ca(s) + __$H_2O$(l) → ________

*2.* __Mg(s) + __$H_2O$(l) → ________

*3.* __Al(s) + __$H_2O$(l) → ________

*4.* __Zn(s) + __HCl(l) → ________

*5.* __Pb(s) + __HCl(l) → ________

*6.* __Cu(s) + __HCl(l) → ________

*7.* __Al(s) + __$NaNO_3$(aq) → ________

*8.* __Al(s) + __$CuCl_2$(aq) → ________

*9.* __Cu(s) + __$AgNO_3$(aq) → ________

**Double replacement reactions**

12. *Test tube no.*

*1.* __NaCl(aq) + __$KNO_3$(aq) → ________

*2.* __NaCl(aq) + __$AgNO_3$(aq) → ________

*3.* __$Na_2CO_3$(aq) + __HCl(aq) → ________

*4.* __NaOH(aq) + __HCl(aq) → ________

*5.* __$BaCl_2$(aq) + __$H_2SO_4$(aq) → ________

*6.* __$Pb(NO_3)_2$(aq) + __$K_2CrO_4$(aq) → ________

*7.* __$Fe(NO_3)_3$(aq) + __NaOH(aq) → ________

*8.* __$Cu(NO_3)_2$(aq) + __NaOH(aq) → ________

## Post-Lab Questions

1. Magnesium metal, Mg, reacts with 0.5 M HCl, but copper metal, Cu, does not. Why?

2. From the following list of chemicals, select two combinations that would lead to a double replacement reaction. Write the complete, balanced equations for the reactions of the chemicals in solution. Use Table 3.2 for guidance.

   $KCl$, $HNO_3$, $AgNO_3$, $PbCl_2$, $Na_2SO_4$

3. You can detect the carbon dioxide, $CO_2$, in your breath by blowing through a straw into an aqueous solution of calcium oxide, CaO. The white precipitate of calcium carbonate, $CaCO_3$, forms. Write the balanced equation for this reaction. This reaction belongs to which class?

4. Fuel cells are important sources of energy for crafts used in the space program. These fuel cells use the reaction of hydrogen and oxygen in a controlled manner to generate electricity and potable water. Write the balanced equation for this reaction. This reaction belongs to which class?

# *Physical Properties of Chemicals: Melting Point, Sublimation, and Boiling Point*

## OBJECTIVES

1. To use melting points and boiling points in identifying substances.
2. To use sublimation as a means of purification.

## BACKGROUND

If you were asked to describe a friend, most likely you would start by identifying particular physical characteristics. You might begin by giving your friend's height, weight, hair color, eye color, or facial features. These characteristics would allow you to single out the individual from a group.

Chemicals also possess distinguishing physical properties which enable their identification. In many circumstances a thorough determination of the physical properties of a given chemical can be used for its identification. If faced with an unknown sample, a chemist may compare the physical properties of the unknown to properties of known substances that are tabulated in the chemical literature; if a match can be made, an identification can be assumed (unless chemical evidence suggests otherwise).

The physical properties most commonly listed in handbooks of chemical data are color, crystal form (if a solid), refractive index (if a liquid), density (discussed in Experiment 2), solubility in various solvents, melting point, sublimation characteristics, and boiling point. When a new compound is isolated or synthesized, these properties almost always accompany the report in the literature.

The transition of a substance from a solid to a liquid to a gas, and the reversal, represent physical changes. In a physical change there is only a change in the form or state of the substance. No chemical bonds break; no alteration in the chemical composition occurs. Water undergoes state changes from ice to liquid water to steam; however, the composition of molecules in all three states remains $H_2O$.

$$\underset{\text{Ice}}{H_2O(s)} \rightleftharpoons \underset{\text{Liquid}}{H_2O(l)} \rightleftharpoons \underset{\text{Steam}}{H_2O(g)}$$

The *melting* or *freezing point* of a substance refers to the temperature at which the solid and liquid states are in equilibrium. The terms are interchangeable and correspond to the same temperature; how the terms are applied depends upon the state the substance is in originally. The melting point is the temperature at equilibrium when starting in the solid state and going to the liquid state. The freezing point is the temperature at equilibrium when starting in the liquid state and going to the solid state.

Melting points of pure substances occur over a very narrow range and are usually quite sharp. The criteria for purity of a solid is the narrowness of the melting point range and the correspondence to the value found in the literature. Impurities will lower the melting point and cause a broadening of the range. For example, pure benzoic acid has a reported melting point of 122.13 °C; benzoic acid with a melting point range of 121–122 °C is considered to be quite pure.

The *boiling point* or *condensation point* of a liquid refers to the temperature when its vapor pressure is equal to the external pressure. If a beaker of liquid is brought to a boil in your laboratory, bubbles of vapor form throughout the liquid; these bubbles rise rapidly to the surface, burst and release vapor to the space above the liquid. In this case, the liquid is in contact with the atmosphere; the normal boiling point of the liquid will be the temperature when the pressure of the vapor is equal to the atmospheric pressure (1 atm or 760 mm Hg). Should the external pressure vary, so will the boiling point. A liquid will boil at a higher temperature when the external pressure is higher and will boil at a lower temperature when the external pressure is reduced. The change in state from a gas to a liquid represents condensation and is the reverse of boiling. The temperature for this change of state is the same as the boiling temperature but is concerned with the approach from the gas phase.

Just as a solid has a characteristic melting point, a liquid has a characteristic boiling point. At one atmosphere, pure water boils at 100 °C, pure ethanol (ethyl alcohol) boils at 78.5 °C, and pure diethyl ether boils at 34.6 °C. The vapor pressure curves shown in Fig. 4.1 illustrate the variation of the vapor pressure of these liquids with temperature. One can use these curves to predict the boiling point at a reduced pressure. For example, diethyl ether has a vapor pressure of 422 mm Hg at 20°C. If the external pressure were reduced to 422 mm Hg, diethyl ether would boil at 20 °C.

*Sublimation* is a process that involves the direct conversion of a solid to a gas without passing through the liquid state. Relatively few solids do this at atmospheric pressure. Some examples are the solid compounds naphthalene (mothballs), caffeine, iodine, and solid carbon dioxide (commercial Dry Ice). Water, on the other hand, sublimes at −10 °C and at 0.001 atm. Sublimation temperatures are not as easily obtained as melting points or boiling points.

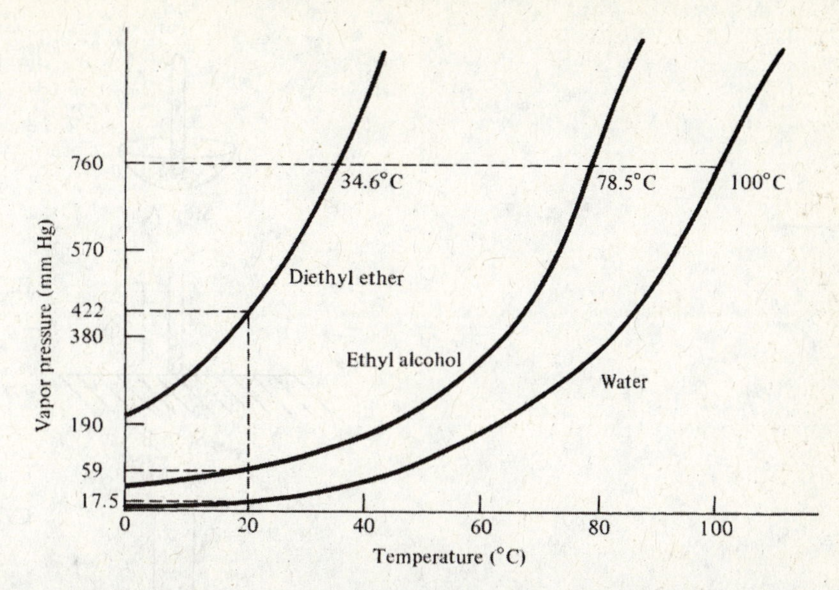


**Figure 4.1**
*Diethyl ether, ethyl alcohol (ethanol), and water vapor pressure* curves.

CAUTION!

**Wear safety glasses throughout this experiment. One should always be prepared for a mishap, such as glass shattering or a liquid suddenly frothing.**

## PROCEDURE

### Melting Point Determination

1. Unknowns are provided by the instructor. Obtain approximately 0.1 g of unknown solid and place it on a small watch glass. Record the number of the unknown on the Report Sheet (1). (The instructor will measure out a 0.1-g sample as a demonstration; take approximately that amount with your spatula.) Carefully crush the solid on a watch glass into a powder with the flat portion of a spatula.

2. Obtain a melting point capillary tube. One end of the tube will be sealed. The tube is packed with solid in the following way:

   **Step A.** Press the open end of the capillary tube vertically into the solid sample (Fig. 4.2 A). A small amount of sample will be forced into the open end of the capillary tube.

   **Step B.** Invert the capillary tube so that the closed end is pointing toward the bench top. Gently tap the end of the tube against the lab bench top (Fig. 4.2 B). Continue tapping until the solid is forced down to the closed end. A sample depth of 5-10 mm is sufficient.

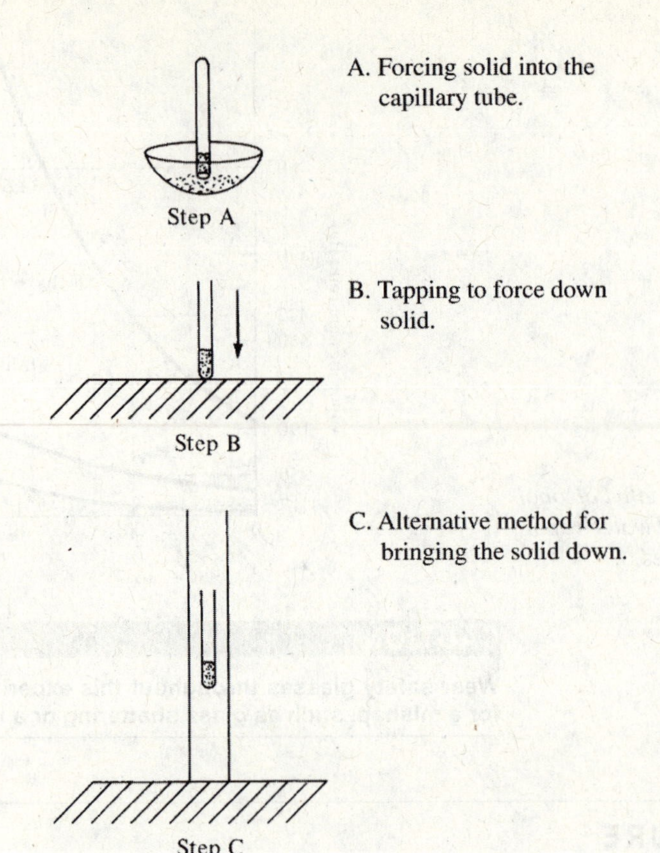
A. Forcing solid into the capillary tube.
Step A
B. Tapping to force down solid.
Step B
C. Alternative method for bringing the solid down.
Step C

**Figure 4.2**
*Packing a capillary tube.*

**Step C.** An alternative method for bringing the solid sample to the closed end uses a piece of glass tubing of approximately 20 to 30 cm. Hold the capillary tube, closed end down, at the top of the glass tubing, held vertically; let the capillary tube drop through the tubing so that it hits the lab bench top. The capillary tube will bounce and bring the solid down. Repeat if necessary (Fig. 4.2 C).

3. The melting point may be determined using either a Thiele tube (Fig. 4.4 A) or a commercial melting point apparatus (Fig. 4.4 B).

   a. A commercial melting point apparatus will be demonstrated by your instructor.

   b. The use of the Thiele tube is as follows:

      (1) Attach the melting point capillary tube to the thermometer by means of a rubber ring. Align the mercury bulb of the thermometer so that the tip of the melting point capillary containing the solid is next to it (Fig. 4.3).

**CAUTION!** 

**The Thiele tube must be completely dry and free of water. Water does not mix with oil. When heated to above 100 °C, the water will boil and cause the oil to bump, leading to the hot oil splashing out of the tube.**

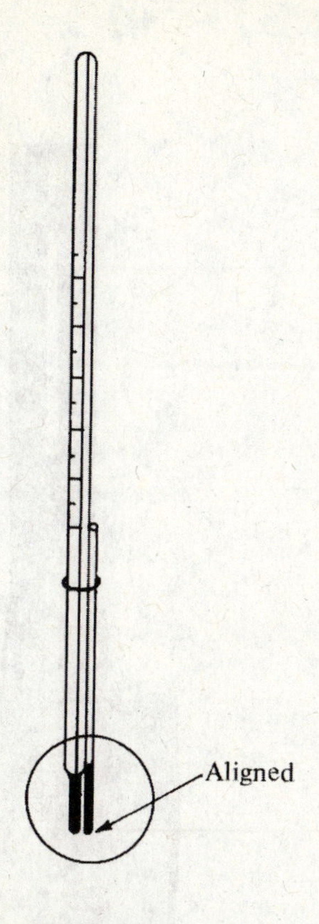


**Figure 4.3**
*Proper alignment of the capillary tube and the mercury bulb.*

(2) Use an extension clamp to support the Thiele tube on a ring stand. Add mineral oil or silicone oil to the Thiele tube, filling to a level above the top of the side arm. Use a thermometer clamp to support the thermometer with the attached melting point capillary tube in the oil. The bulb and capillary tube should be immersed in the oil; keep the rubber ring and open end of the capillary tube out of the oil (Fig. 4.4 A).

(3) Heat the arm of the Thiele tube very slowly with a Bunsen burner flame. Use a small flame and gently move the burner along the arm of the Thiele tube.

(4) You should position yourself so that you can follow the rise of the mercury in the thermometer as well as observe the solid in the capillary tube. Record the temperature when the solid begins to liquefy (the solid will appear to shrink). This is the temperature at which the solid starts to melt. Record the temperature when the solid is completely a liquid. This is the temperature at which the melting ends. Express these readings as a melting point range (2).

(5) Identify the solid by comparing the melting point with those listed in Table 4.1 for different solids (3).

4. Do as many melting point determinations as your instructor may require. Just remember to use a new melting point capillary tube for each melting point determination.

5. Dispose of the solids as directed by your instructor.

**Table 4.1** *Melting Points of Selected Solids*

| *Solids* | *Melting Point (°C)* |
|---|---|
| Acetamide | 82 |
| Acetanilide | 114 |
| Benzophenone | 48 |
| Benzoic acid | 122 |
| Biphenyl | 70 |
| Lauric acid | 43 |
| Naphthalene | 80 |
| Stearic acid | 70 |

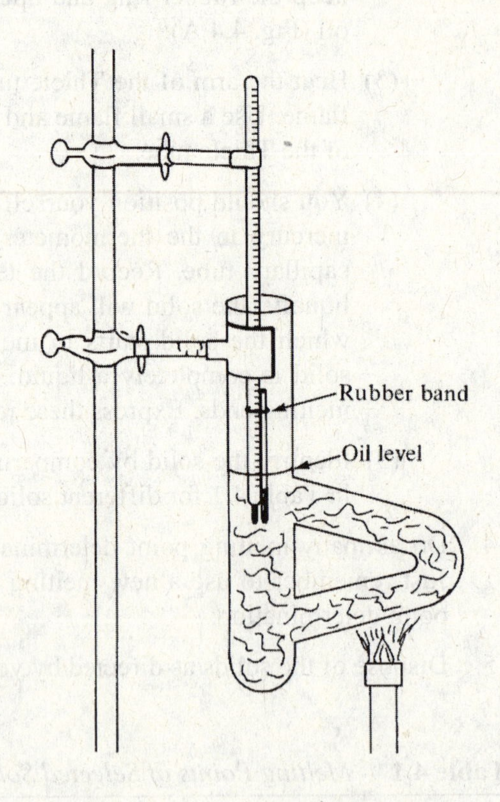


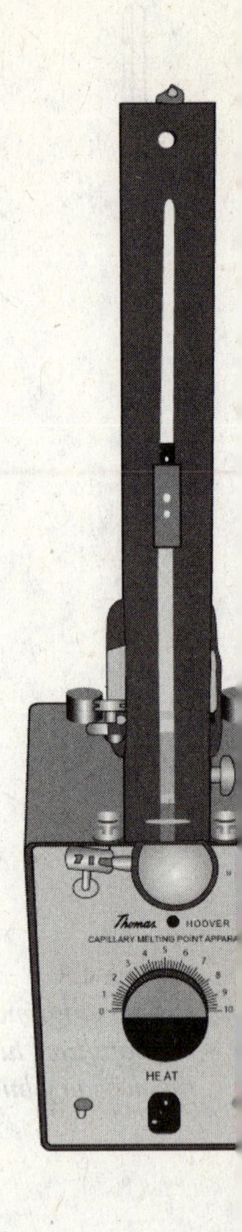


**Figure 4.4**
***A.*** *Thiele tube apparatus.*
***B*** *A commercial apparatus.*

A. B.

**Purification of Naphthalene by Sublimation**

1. Place approximately 0.5 g of impure naphthalene into a 100-mL beaker. (Your instructor will measure out 0.5 g of sample; with a spatula take an amount which approximates this quantity.)
2. Into the 100-mL beaker place a smaller 50-mL beaker. Fill the smaller beaker halfway with ice cubes or ice chips. Place the assembled beakers on a wire gauze supported by a ring clamp (Fig. 4.5).

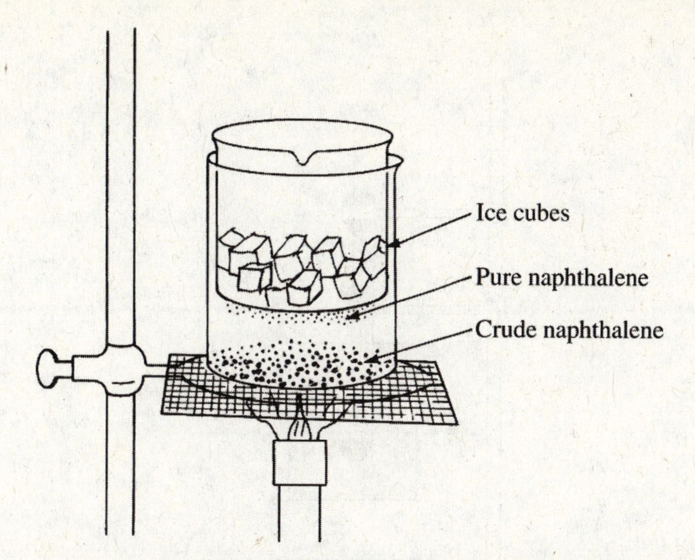


**Figure 4.5**
*Set-up for sublimation of naphthalene.*

3. Using a small Bunsen burner flame, gently heat the bottom of the 100-mL beaker by passing the flame back and forth beneath the beaker.
4. You will see solid flakes of naphthalene collect on the bottom of the 50-mL beaker. When a sufficient amount of solid has collected, turn off the burner.
5. Pour off the ice water from the 50-mL beaker and carefully scrape the flakes of naphthalene onto a piece of filter paper with a spatula.
6. Take the melting point of the pure naphthalene and compare it to the value listed in Table 4.1 (4).
7. Dispose of the crude and pure naphthalene as directed by your instructor.

**CAUTION!**

**The chemicals used for boiling point determinations are flammable. Be sure all Bunsen burner flames are extinguished before completing this part of the experiment.**

**Boiling Point Determination**

1. Obtain from your instructor an unknown liquid and record its number on the Report Sheet (5).
2. Clamp a clean, dry test tube (13 × 100 mm) onto a ring stand. Add to the test tube approximately 3 mL of the unknown liquid and two small boiling chips. Lower the test tube into a 250-mL beaker which contains 100 mL of water and two boiling chips. Adjust the depth of the test tube so that the unknown liquid is below the water level of the water bath (Fig. 4.6).

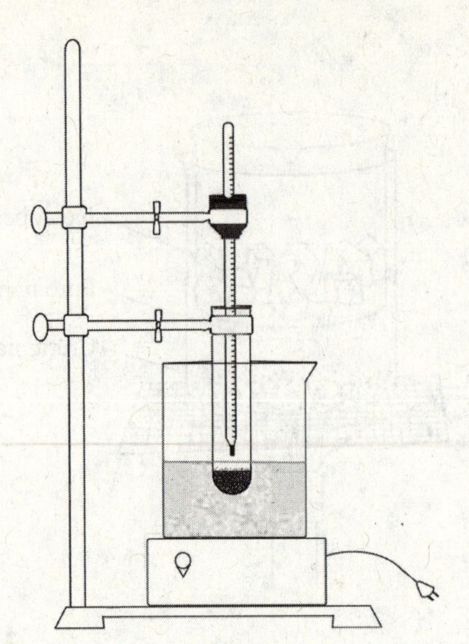

**Figure 4.6**
*Set-up for determining the boiling point.*

3. Pass a thermometer through a neoprene adapter and secure it with a thermometer clamp; lower it into the test tube. Adjust the thermometer so that it is approximately 1 cm above the surface of the unknown liquid.
4. A piece of aluminum foil can be used to cover the mouth of the test tube.

**CAUTION!**

**Be certain that the test-tube mouth has an opening so pressure does not build up; the system should not be closed.**

5. Gradually heat the water in the beaker with a hot plate and watch for changes in temperature. As the liquid begins to boil, the temperature above the liquid will rise. When the temperature no longer rises but remains constant, record the temperature to the nearest 0.1 °C (6). This is the observed boiling point. From the list in Table 4.2, identify your unknown liquid by matching your observed boiling point with the compound whose boiling point best corresponds (7).
6. Do as many boiling point determinations as required by your instructor.
7. Dispose of the liquid as directed by your instructor.

**Table 4.2** *Boiling Points of Selected Liquids*

| *Liquid* | *Boiling Point (°C at 1 atm)* | *Use* |
|---|---|---|
| Acetone | 56 | Solvent; paint remover |
| Cyclohexane | 81 | Solvent for lacquers and resins |
| Ethyl acetate | 77 | Solvent for airplane dopes; artificial fruit essence |
| Hexane | 69 | Liquid in thermometers with blue or red dye |
| Methanol (methyl alcohol) | 65 | Solvent; radiator antifreeze |
| 1-Propanol | 97 | Solvent |
| 2-Propanol (isopropyl alcohol) | 83 | Solvent for shellac; essential oils; body rubs |

## CHEMICALS AND EQUIPMENT

1. Aluminum foil
2. Boiling chips
3. Bunsen burner
4. Hot plate
5. Commercial melting point apparatus (if available)
6. Melting point capillary tubes
7. Rubber rings
8. Thiele tube melting point apparatus
9. Thermometer clamp
10. Glass tubing
11. Acetamide
12. Acetanilide
13. Acetone
14. Adipic acid
15. Benzophenone
16. Benzoic acid
17. Cyclohexane
18. *p*-Dichlorobenzene
19. Ethyl acetate
20. Hexane
21. Methanol (methyl alcohol)
22. Naphthalene, pure
23. Naphthalene, impure
24. 1-Propanol
25. 2-Propanol (isopropyl alcohol)
26. Stearic acid
27. Test tube (13 × 100 mm)
28. 50-mL beaker
29. 100-mL beaker

name (print) | date (of lab meeting) | grade

course/section | partner's name (if applicable)

4 **EXPERIMENT 4: PHYSICAL PROPERTIES OF CHEMICALS**

# *Prelab Questions*

**A. Safety concerns.**

1. Why is it necessary to wear safety glasses for this experiment?

2. What is the problem in using Bunsen burners in this experiment?

3. Why do you avoid a closed system?

**B. Basic principles.**

1. Why is the transition of water from the solid to a liquid a physical change and not a chemical change?

2. Refer to Fig. 4.1. What happens to the boiling points of the three liquids as the external pressure is reduced?

3. What are the criteria for purity of a solid?

name (print) ______ date (of lab meeting) ______ grade ______

course/section ______ partner's name (if applicable) ______



# *Report Sheet*

**Melting point determination**

| | Trial No. 1 | Trial No. 2 |
|---|---|---|
| 1. Code number of unknown | ______ | ______ |
| 2. Melting point range (°C) | | |
| Temperature melting begins | ______°C | ______°C |
| Temperature melting ends | ______°C | ______°C |
| 3. Identification of unknown | ______ | ______ |
| **Purification of naphthalene by sublimation** | | |
| 4. Melting point range (°C) | | |
| Temperature melting begins | ______°C | ______°C |
| Temperature melting ends | ______°C | ______°C |
| **Boiling point determination** | | |
| 5. Unknown number | ______ | ______ |
| 6. Observed boiling point | ______°C | ______°C |
| 7. Identification of unknown | ______ | ______ |

## Post-Lab Questions

1. A student did a melting point determination for a sample of acetanilide and found a melting point of 113–114 °C. What conclusion can the student draw about the sample?

2. A student in New York City carried out a boiling point determination for cyclohexane (b.p. 81 °C) according to the procedure in this laboratory manual. Will this student's observed boiling point be the same as the value obtained by another student in Denver, Colorado (nicknamed the "Mile-High City")? Will it be lower or higher? Explain your conclusion.

3. An ice chest containing solid carbon dioxide (Dry Ice) was left open and the inside warmed to room temperature. When examined later, there was no solid and no liquid on the bottom of the chest. What happened to everything?

4. Cocaine is a white solid which melts at 98 °C when pure. A forensic chemist working for the New York Police Department has a white solid believed to be cocaine. What can the chemist do to quickly determine whether the sample is cocaine and whether it is pure or a mixture?

# *Factors Affecting Rate of Reactions*

## OBJECTIVES

1. To investigate the relationship between the rate and the nature of reactants.
2. To measure the rate of reaction as a function of concentration.
3. To demonstrate the effect of temperature on the rate of reaction.
4. To investigate the effect of surface area and the effect of a catalyst on the rate of reaction.

## BACKGROUND

Did you ever wonder why some chemical reactions take place rapidly, while others are very slow? An antacid tablet can neutralize stomach acid (HCl) rapidly. The reaction of hydrogen and oxygen with each other to form water is extremely slow; a tank containing a mixture of $H_2$ and $O_2$ shows no measurable change even after many years, but should a spark occur, the whole tank would explode violently. In this experiment we will try to answer why there are these differences.

The study of rates of reactions is called *chemical kinetics*. The *rate of reaction* is the change in concentration of a reactant (or product) per unit time. For example, if in the reaction

$$2HCl(aq) + CaCO_3(s) \rightarrow CaCl_2(aq) + H_2O(l) + CO_2(g)$$

suppose we monitor the evolution of $CO_2$, and we find that 4.4 g of carbon dioxide gas was produced in 10 min. Since 4.4 g corresponds to 0.1 moles of $CO_2$, then the rate of reaction is 0.01 moles $CO_2$/min. On the other hand, if we monitor the HCl concentration, we may find that at the beginning we had 0.6 M HCl and after 10 min. the concentration of HCl was 0.4 M HCl. This means that we used up 0.2 M HCl in 10 min. Thus the rate of reaction is 0.02 moles HCl/L-min. From the above we can see that when describing the rate of reaction, it is not sufficient to give a number, we have to specify the units and also the reactant (or product) we monitored. In order that a reaction should take place, molecules or ions must first collide. Not every collision yields a reaction. In many collisions, the molecules simply bounce apart without reacting. A collision that results in a reaction is called an *effective collision.* The minimum energy necessary for the reaction to happen is called *the activation energy* (Fig. 5.1). In this energy diagram we see that the rate of reaction depends on this activation energy.

The lower the activation energy the faster the rate of reaction; the higher the activation energy the slower the reaction. This is true for both exothermic and endothermic reactions.

A number of factors affect the rates of reactions. In our experiments we will see how these affect the rates of reactions.

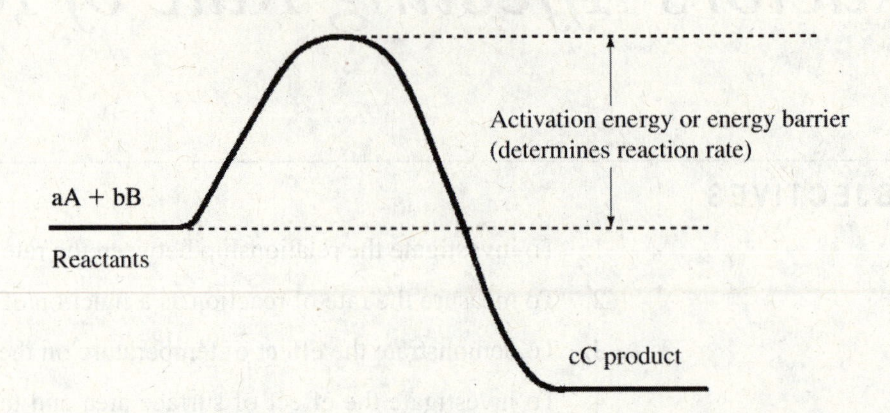


**Figure 5.1**
*Energy diagram for a typical reaction.*

A. **Nature of the reactants.** Some compounds are more reactive than others. In general, reactions that take place between ions in aqueous solutions are rapid. Reactions between covalent molecules are much slower.

B. **Concentration.** In most reactions, the rate increases when the concentration of either or both reactants is increased. This is understandable on the basis of the collision theory. If we double the concentration of a reactant, it will collide in each second twice as many times with the second reactant as before. Since the rate of reaction depends on the number of effective collisions per second, the rate is doubled (Fig. 5.2).

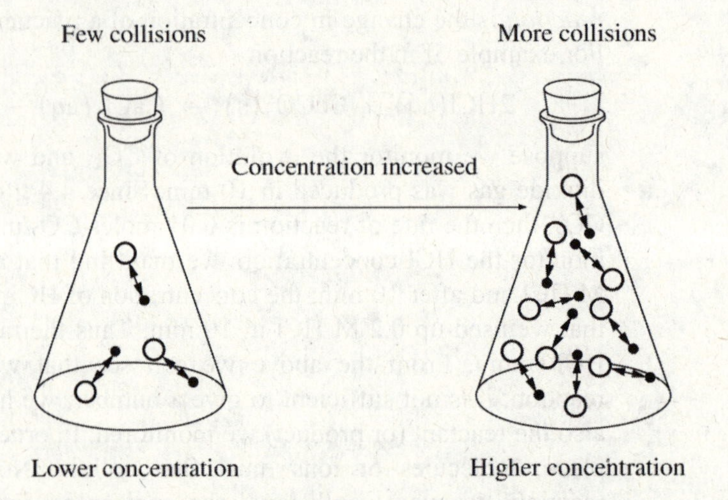


**Figure 5.2**
*Concentration affecting the rate of reaction.*

C. **Surface area.** If one of the reactants is a solid, the molecules of the second reactant can collide only with the surface of the solid. Thus the surface area of the solid is in effect its concentration. An increase in the surface area of the solid (by grinding to a powder in a mortar) will increase the rate of reaction.

D. **Temperature.** Increasing the temperature makes the reactants more energetic than before. This means that more molecules will have energy equal to or greater than the activation energy. Thus one expects an increase in the rate of reaction with increasing temperature. As a rule of thumb, every time the temperature goes up by 10 °C, the rate of reaction doubles. This rule is far from exact, but it applies to many reactions.

E. **Catalyst.** Any substance that increases the rate of reaction without itself being used up in the process is called a *catalyst*. A catalyst increases the rate of reaction by lowering the activation energy (Fig. 5.3). Thus many more molecules can cross the energy barrier (activation energy) in the presence of a catalyst than in its absence. Almost all the chemical reactions in our bodies are catalyzed by specific catalysts called enzymes.

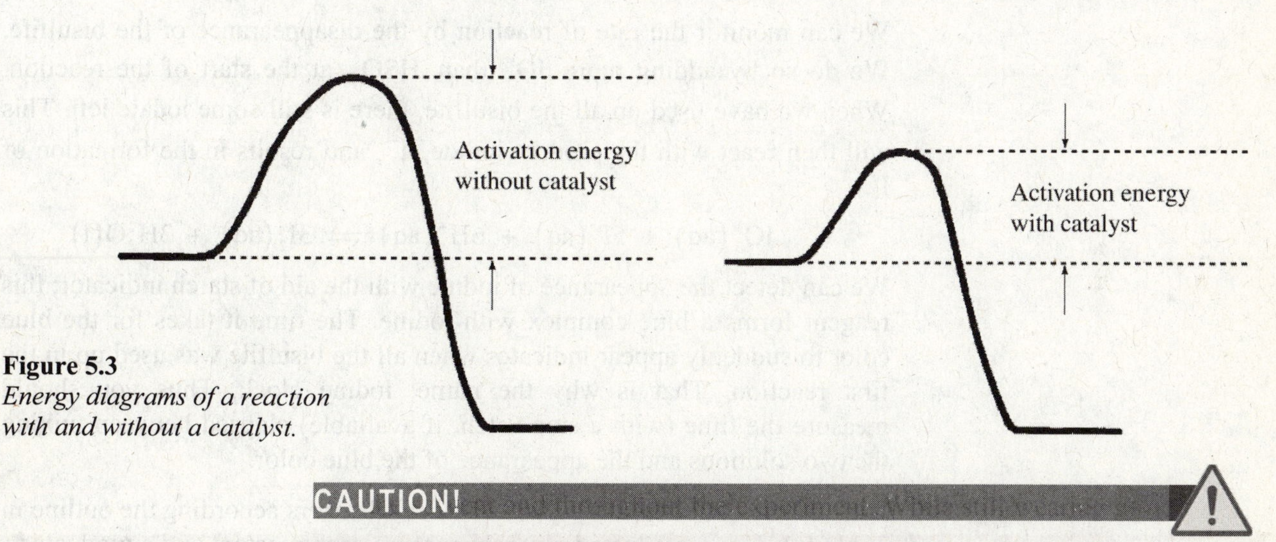


**Figure 5.3**
*Energy diagrams of a reaction with and without a catalyst.*

**CAUTION!**

**Wear safety glasses throughout this experiment. Handle acids with care. Use gloves. Wash any acid spill thoroughly with water for at least 15 min.**

## PROCEDURE

**A. Nature of reactants.**

1. Label five (10 × 75 mm) test tubes 1 through 5.
2. Place in each test tube one 1-cm polished strip of magnesium ribbon. Add 1 mL of acid to each test tube as follows: no. 1) 3 M $H_2SO_4$; no. 2) 6 M HCl; no. 3) 6 M $HNO_3$; no. 4) 2 M $H_3PO_4$; and no. 5) 6 M $CH_3COOH$.
3. The reaction will convert the magnesium ribbon to the corresponding salts with the liberation of hydrogen gas. You can assess the rate of reaction qualitatively, by observing the speed with which the gas is liberated (bubbling) and/or by noticing the time of disappearance of the magnesium ribbon. **Do all of the reactions in the five test tubes at the same time**; assess the rates of reaction; then list, in decreasing order, the rates of reaction of magnesium with the various acids on your Report Sheet (1).

4. Place 1 mL 6 M HCl in three test tubes (10 × 75 mm) numbered 1 through 3. Add 1-cm polished strips of magnesium ribbon to test tube no. 1, zinc ribbon to test tube no. 2, and copper ribbon to test tube no. 3. **Do all of the reactions in the three test tubes at the same time**; assess the rates of reaction of the three metals by the speed of evolution of $H_2$ gas; then list, in decreasing order, the rates of reaction of the metals with the acid on your Report Sheet (2).

**B. Concentration**. The *iodine clock* reaction is a convenient reaction for observing concentration effects. The reaction is between potassium iodate, $KIO_3$, and sodium bisulfite, $NaHSO_3$; the net ionic reaction is given by the following equation.

$$IO_3^-(aq) + 3HSO_3^-(aq) \rightleftharpoons I^-(aq) + 3SO_4^{2-}(aq) + 3H^+(aq)$$

We can monitor the rate of reaction by the disappearance of the bisulfite. We do so by adding more $IO_3^-$ than $HSO_3^-$ at the start of the reaction. When we have used up all the bisulfite, there is still some iodate left. This will then react with the product iodide, $I^-$, and results in the formation of $I_2$.

$$IO_3^-(aq) + 5I^-(aq) + 6H^+(aq) \rightleftharpoons 3I_2(aq) + 3H_2O(l)$$

We can detect the appearance of iodine with the aid of starch indicator; this reagent forms a blue complex with iodine. The time it takes for the blue color to suddenly appear indicates when all the bisulfite was used up in the first reaction. That is why the name: iodine clock. Thus you should measure the time (with a stopwatch, if available) elapsed between mixing the two solutions and the appearance of the blue color.

1. Place the reactants in two separate 150-mL beakers according the outline in Table 5.1. Use a graduated pipet to measure each reactant and a graduated cylinder to measure the water.

**Table 5.1** *Reactant Concentration and Rate of Reaction*

| Beaker A | | | | Beaker B | |
|---|---|---|---|---|---|
| *Trial* | *0.1 M $KIO_3$* | *Starch* | *Water* | *0.01 M $NaHSO_3$* | *Water* |
| 1 | 2.0 mL | 2 mL | 46 mL | 5 mL | 45 mL |
| 2 | 4.0 mL | 2 mL | 44 mL | 5 mL | 45 mL |
| 3 | 6.0 mL | 2 mL | 42 mL | 5 mL | 45 mL |

2. **At exactly same time**, pour the contents of Beaker A and Beaker B into a third beaker and time the appearance of the blue color.

3. Repeat the experiment with the other two trial concentrations. Record your data on the Report Sheet (3).

**C. Surface area.**

1. Using a large mortar and pestle, crush and pulverize about 0.5 g of marble chips.

2. Place the crushed marble chips into one large test tube (16 × 125 mm) and 0.5 gram of marble chip chunks into another.

3. Add 2 mL of 6 M HCl to each test tube and note the speed of bubbling of the $CO_2$ gas. Record your data on the Report Sheet (4).

**D. Temperature.**

1. Add 5 mL 6 M HCl to three clean test tubes (16 × 125 mm) numbered 1 through 3.

2. Place test tube no. 1 in an ice bath (4 °C); test tube no. 2 in a beaker containing warm water (50 °C); and test tube no. 3 in a beaker with tap water (20 °C). Wait 5 min. To each test tube add a piece of zinc ribbon (1 cm × 0.5 cm × 0.5 mm). Note the time you added the zinc metal. Finally, note the time when the bubbling of gas stops in each test tube and the zinc metal disappears. Record the time of reaction (time of the disappearance of the zinc minus the time of the start of the reaction) on your Report Sheet (5).

**E. Catalyst.**

1. Add 2 mL of 3% $H_2O_2$ solution to two clean test tubes (16 × 125 mm). The evolution of oxygen bubbles will indicate whether the hydrogen peroxide decomposed. Note if anything happens.

2. Now add a few grains of manganese dioxide, $MnO_2$, to one of the test tubes. Note if there is evolution of oxygen. Record your data on the Report Sheet (6).

## CHEMICALS AND EQUIPMENT

1. Mortar and pestle
2. 10-mL graduated pipet
3. 5-mL volumetric pipet
4. Magnesium ribbon
5. Zinc ribbon
6. Copper ribbon
7. 3 M $H_2SO_4$
8. 6 M HCl
9. 6 M $HNO_3$
10. 2 M $H_3PO_4$
11. 6 M $CH_3COOH$
12. 0.1 M $KIO_3$
13. 0.01 M $NaHSO_3$
14. Starch indicator
15. Marble chips
16. 3% hydrogen peroxide
17. Manganese dioxide
18. Test tubes (10 × 75 mm)
19. Test tubes (16 × 125 mm)

name (print) ______________________ date (of lab meeting) ______________________ grade ______________________

course/section ______________________ partner's name (if applicable) ______________________

**5** **EXPERIMENT 5: FACTORS AFFECTING RATE OF REACTION**

# *Prelab Questions*

**A. Safety concerns.**

1. How do you protect your eyes against foreign objects?

2. What do you do to protect your hands from acid burns?

3. You get some acid on your bare hands. What do you immediately do?

**B. Basic principles.**

1. You have heartburn and wish to get relief as quickly as possible. One antacid comes in the form of a compressed pill; another is in the form of loose powder. Which form would give faster relief for your heartburn, considering that they contained the same ingredients and were taken in equal amounts? Why?

2. You increase the temperature of a reaction significantly. What do you expect to happen?

name (print) ______ date (of lab meeting) ______ grade ______

course/section ______ partner's name (if applicable) ______

5 **EXPERIMENT 5: FACTORS AFFECTING RATE OF REACTION**

# *Report Sheet*

A. **Nature of reactants** **Name of the acid**

1. Fastest reaction ______

______

______

______

Slowest reaction ______

**Name of the metal**

2. Fastest reaction ______

______

Slowest reaction ______

B. **Effect of concentration**

Trial no. Time

1 ______

2 ______

3 ______

C. **Surface area**

Fast reaction ______

Slow reaction ______

D. **Effect of temperature**

| Trial no. at | 4°C | 20°C | 50°C |
| --- | --- | --- | --- |
| Reaction time | ______ | ______ | ______ |

E. **Catalyst** **Observation**

No $MnO_2$ ________________________________

With $MnO_2$ ________________________________

## Post-Lab Questions

1. When manganese dioxide was added to the hydrogen peroxide solution, bubbling was observed when there was none before. What role did the manganese dioxide play in the reaction?

2. If in the reaction between magnesium and HCl, we would have used 3 M HCl rather than 6 M HCl, would the rate of reaction increase, decrease, or remain the same? Explain you answer.

3. If in the same reaction we would have used globular chunks of magnesium instead of ribbon (both having the same weight), would the rate of reaction increase, decrease or remain the same? What would be affected (if any) by this change?

4. In the **PRE-LAB** questions, you were asked about the effectiveness of a powdered antacid versus one available as a pill for the relief of heartburn. Is there anything in this experiment that would support an answer that a powder might be faster than a pill?

5. Assume we do a reaction of zinc with 6 M HCl at room temperature, 20 °C. How much faster will these two chemicals react at 40 °C (see part D)?

EXPERIMENT 6

# The Law of Chemical Equilibrium and The Le Chatelier Principle

## OBJECTIVES

1. To study chemical equilibria.
2. To investigate the effects of changing concentrations and changing temperature in equilibrium reactions.

## BACKGROUND

A chemist asks two important questions for every chemical reaction. How much product is produced? How fast is it produced? The first question involves chemical equilibrium. The second question concerns chemical kinetics. In Experiment 5 we looked at kinetics. In this experiment, we will investigate chemical equilibria.

Some reactions are irreversible and they go to completion (100% yield). When you ignite methane gas in a gas burner in the presence of air (oxygen), methane burns completely and forms carbon dioxide and water.

$$CH_4(g) + 2O_2(g) \rightarrow CO_2(g) + 2H_2O(g)$$

Other reactions do not go to completion. They are reversible. In such cases, the reaction can go in either direction: forward or backward. For example, the reaction

$$\underset{\text{yellow}}{Fe^{3+}(aq)} + SCN^-(aq) \rightleftharpoons \underset{\text{red}}{FeSCN^{2+}(aq)}$$

is often used to illustrate reversible reactions. This reaction is used because it is easy to observe the progress of the reaction visually. The yellow iron(III), $Fe^{3+}$, ion reacts with thiocyanate, $SCN^-$, ion to form a deep red complex ion, $FeSCN^{2+}$. This is the forward reaction. At the same time the complex red ion also decomposes and forms the yellow iron(III) ion and thiocyanate ion. This is the backward (reverse) reaction. At the beginning when we mix iron(III) ion and thiocyanate ion the rate of the forward reaction is at a maximum and as time goes on, this rate decreases, because we have less and less iron(III) and thiocyanate to react. On the other hand, the rate of the reverse reaction (which began at zero) gradually increases. Eventually the two rates become equal. When this point is reached, we call the process a *dynamic equilibrium* or just *equilibrium*. When in equilibrium at a particular temperature, a reaction mixture obeys the *Law of Chemical Equilibrium*. This Law imposes a condition on the concentration of reactants and products expressed in the equilibrium constant (K). For the above reaction between iron(III) and thiocyanate ions, the equilibrium constant can be written as

$$K = \frac{[FeSCN^{2+}]}{[Fe^{3+}][SCN^-]}$$

or in general

$$K = \frac{[products]}{[reactants]}$$

The brackets, [ ], indicate concentration, in moles/L, at equilibrium. As the name implies the *equilibrium constant* is a constant at a set temperature for a particular reaction. Its magnitude tells if a reaction goes to completion or if it is far from completion (reversible reaction). A number much smaller than 1 for K indicates that at equilibrium only a few molecules of products are formed, meaning the mixture consists mainly of reactants. We say that the equilibrium lies far to the left. On the other hand, a complete reaction (100% yield), would have a very large number for an equilibrium constant. In this case, obviously the equilibrium lies far to the right. The above reaction between iron(III) ion and thiocyanate ion has an equilibrium constant of 207, indicating that the equilibrium lies to the right but does not go to completion. Thus at equilibrium, both reactants and product are present, albeit the products far outnumber the reactants.

The Law of Chemical Equilibrium is based on the constancy of the equilibrium constant. This means that if one disturbs the equilibrium, for example by adding more reactant molecules, there will be an increase in the number of product molecules in order to maintain the product/reactant ratio unchanged and thus, preserving the numerical value of the equilibrium constant. The *Le Chatelier Principle* expresses this as follows: *If an external stress is applied to a system in equilibrium, the system reacts in such a way as to partially relieve the stress.* In our present experiment, we will demonstrate the Le Chatelier Principle in two ways. We will disturb the equilibrium, (a) by changing the concentration of a product or reactant, and (b) by changing the temperature.

**A. Concentration effects.**

1. In the first experiment, we add ammonia to a pale blue copper(II) sulfate solution. The ionic reaction is

$$\underset{\text{pale blue}}{Cu(H_2O)_4^{2+}(aq)} + \underset{\text{colorless}}{4NH_3(aq)} \rightleftharpoons \underset{\text{(color?)}}{Cu(NH_3)_4^{2+}(aq)} + 4H_2O(l)$$

A change in the color indicates the copper-ammonia complex formation. Adding a strong acid, HCl, to this equilibrium causes the ammonia, $NH_3$, to react with the acid:

$$NH_3(aq) + H^+(aq) \rightleftharpoons NH_4^+(aq)$$

Thus we removed some reactant molecules from the equilibrium mixture. As a result we expect the equilibrium to shift to the left, reforming hydrated copper(II) ions with the reappearance of the pale blue color.

2. In the second reaction, we demonstrate the common ion effect. When we have a mixture of $H_2PO_4^-/HPO_4^{2-}$ solution, the following equilibrium exists:

$$H_2PO_4^-(aq) + H_2O \rightleftharpoons H_3O^+(aq) + HPO_4^{2-}(aq)$$

If we add a few drops of aqueous HCl to the solution, we added a common ion, $H_3O^+$, that already was present in the equilibrium mixture. We expect, on the basis of the Le Chatelier Principle, that the equilibrium will shift to the left. Thus the solution will become less acidic.

3. In the iron(III)-thiocyanate reaction,

$$\underset{\text{yellow}}{Fe^{3+}(aq)} + 3Cl^-(aq) + K^+(aq) + \underset{\text{colorless}}{SCN^-(aq)} \rightleftharpoons$$

$$\underset{\text{red}}{Fe(SCN)^{2+}(aq)} + 3Cl^-(aq) + \underset{\text{colorless}}{K^+(aq)}$$

the chloride and potassium ions are spectator ions. Nevertheless, their concentration may also influence the equilibrium. For example, when the chloride ions are in excess, the yellow color of the $Fe^{3+}$ will disappear with the formation of a colorless $FeCl_4^-$ complex:

$$\underset{\text{yellow}}{Fe^{3+}(aq)} + 4Cl^-(aq) \rightleftharpoons \underset{\text{colorless}}{FeCl_4^-(aq)}$$

**B. Temperature effects.**

1. Most reactions are accompanied by some energy changes. Frequently, the energy is in the form of heat. We talk of endothermic reactions if heat is consumed during the reaction. In endothermic reactions, we can consider heat as one of the reactants. Conversely, heat is evolved in an exothermic reaction, and we can consider heat as one of the products. Therefore, if we heat an equilibrium mixture of an endothermic reaction, it will behave as if we added one of its reactants (heat) and the equilibrium will shift to the right. If we heat the equilibrium mixture of an exothermic reaction, the equilibrium will shift to the left. We will demonstrate the effect of temperature on the reaction

$$Co(H_2O)_6^{2+}(aq) + 4Cl^- \rightleftharpoons CoCl_4^{2-}(aq) + 6H_2O(l)$$

You will observe a change in the color depending whether the equilibrium was established at room temperature or at 100°C (in boiling water). From the color change, you should be able to tell if the reaction was endothermic or exothermic.

## PROCEDURE

### Concentration Effects

1. Place 20 drops (about 1 mL) of 0.1 M $CuSO_4$ solution into a clean and dry test tube (13 × 100 mm). Add (dropwise) 1 M $NH_3$ solution, mixing the contents after each drop. Continue to add until the color changes. Note the new color and the number of drops of 1 M ammonia added and record it on your Report Sheet (1). To the equilibrium mixture thus obtained, add (dropwise, counting the number of drops added) 1 M HCl solution until the color changes back to pale blue. Report your observations on your Report Sheet (2).
2. Place 2 mL of $H_2PO_4^-/HPO_4^{2-}$ solution into a clean and dry test tube (13 × 100 mm). Using red and blue litmus papers, test to see whether the solution is acidic or basic. Record your findings on your Report Sheet (3). Add a drop of 1 M HCl to the litmus paper. Record your observation on the Report Sheet (4). Add one drop of 1 M HCl solution to the test tube. Mix it and test it with litmus paper. Record your observation on the Report Sheet (5).
3. Prepare a stock solution by adding 1 mL of 0.1 M iron(III) chloride, $FeCl_3$, and 1 mL of 0.1 M potassium thiocyanate, KSCN, to 50 mL distilled water in a 100-mL beaker. Set up four clean and dry test tubes (13 × 100 mm) and label them nos. 1, 2, 3, and 4. To each test tube add about 2 mL of the stock equilibrium mixture you just prepared. Use the test tube no. 1 as a standard to which you can compare the color of the other solutions. To test tube no. 2, add 10 drops of 0.1 M iron(III) chloride solution; to test tube no. 3, add 10 drops of 0.1 M KSCN solution. To test tube no. 4, add five drops of saturated NaCl solution. Observe the color in each test tube and record your observations on the Report Sheet (6) and (7).

### Temperature Effects

1. Set up two clean and dry test tubes (13 × 100 mm). Label them nos. 1 and 2. Prepare a boiling water bath by heating a 400-mL beaker containing about 200 mL water to a boil.

**CAUTION!**

**Concentrated HCl is toxic and can cause skin burns. Wear gloves when dispensing. Do not allow skin contact. If you do come into contact with the acid, immediately wash the exposed area with plenty of water for at least 15 min. Do not inhale the HCl vapors. Dispense in the hood.**

2. Place 5 drops of 1 M $CoCl_2$ solution in test tube no. 1. Add concentrated HCl dropwise until a color change occurs. Record your observation on the Report Sheet (8).
3. Place 1 mL $CoCl_2$ solution in test tube no. 2. Note the color. Immerse the test tube into the boiling water bath. Report your observations on the Report Sheet (9) and (10).

## CHEMICALS AND EQUIPMENT

1. 0.1 M $CuSO_4$
2. 1 M $NH_3$
3. 1 M HCl
4. Saturated NaCl
5. Concentrated HCl
6. 0.1 M KSCN
7. 0.1 M $FeCl_3$
8. 1 M $CoCl_2$
9. $H_2PO_4^-$ / $HPO_4^{2-}$ solution
10. Litmus paper
11. Test tube (13 × 100 mm)

name (print) | date (of lab meeting) | grade

course/section | partner's name (if applicable)

6

# EXPERIMENT 6: THE LAW OF CHEMICAL EQUILIBRIUM AND THE LE CHATELIER PRINCIPLE

## *Prelab Questions*

**A. Safety concerns.**

1. What are some of the problems to avoid when working with concentrated HCl? How do you work with concentrated HCl in a safe way?

**B. Basic principles.**

1. For the reaction $NH_3(aq) + H^+(aq) \rightleftharpoons NH_4^+(aq)$ at 20°C, the equilibrium constant is calculated to be, $K = 4.5 \times 10^8$.
   a. To which side of the equilibrium is the reaction?

   b. If the reaction is exothermic and the temperature was increased to 40°C, would the equilibrium concentration of $NH_3$ increase, remain the same, or decrease?

2. If the reaction between iron(III) ion and thiocyanate ion yielded an equilibrium concentration of 0.2 M for each of these ions, calculate the equilibrium concentration of the red iron-thiocianate complex?
   (Hint: the equilibrium constant for the reaction is in the **Background**.)

name (print) date (of lab meeting) grade

course/section partner's name (if applicable)

**6** **EXPERIMENT 6: THE LAW OF CHEMICAL EQUILIBRIUM AND THE LE CHATELIER PRINCIPLE**

# *Report Sheet*

1. What is the color of the copper-ammonia complex? ________

   How many drops of 1 M ammonia did you add to cause a change in color? ________

2. How many drops of 1 M HCl did you add to cause a change in color back to pale blue? ________

3. Testing the phosphate solution, what was the color of the red litmus paper? ________

   What was the color of the blue litmus paper? ________

4. Testing the 1 M HCl solution, what was the color of the red litmus paper? ________

   What was the color of the blue litmus paper? ________

5. After adding one drop of 1 M HCl to the phosphate solution, and testing it with litmus paper, what was the color of the red litmus paper? ________

   What was the color of the blue litmus paper? ________

   Was your phosphate solution acidic, basic or neutral,

   a. before the addition of HCl? ________

   b. after the addition of HCl? ________

   Was your solution after adding HCl acidic, basic or neutral? ________

6. Compare the colors in each of the test tubes containing the iron(III) chloride-thiocyanate mixtures:

   no. 1 ________

   no. 2 ________

   no. 3 ________

   no. 4 ________

7. In which direction did the equilibrium shift in test tube

   no. 2 ____________________

   no. 3 ____________________

   no. 4 ____________________

8. What is the color of the $CoCl_2$ solution

   a. before the addition of HCl? ____________

   b. after the addition of HCl? ____________

9. What is the color of $CoCl_2$ solution

   a. at room temperature? ____________

   b. at boiling water temperature? ____________

10. In which direction did the equilibrium shift upon heating? ____________

11. From the above shift, determine if the reaction was exothermic or endothermic. ____________

## Post-Lab Questions

1. Consider the reaction: $Fe^{3+}(aq) + 4Cl^{-}(aq) \rightleftharpoons FeCl_4^{-}(aq)$

   yellow colorless

   What would happen to the color of a dilute solution containing $FeCl_4^{-}$ if

   a. we added a solution containing silver ion? (Silver ion reacts with chloride ion in solution to form the precipitate AgCl.)

   b. we added concentrated HCl?

2. Adding HCl to the $H_2PO_4^{-}/HPO_4^{2-}$ mixture causes a common ion effect. Explain why.

3. The manufacture of ammonia from nitrogen and hydrogen (the Haber process) is an exothermic reaction. Which temperature would give a greater yield of ammonia: room temperature or 100 °C? Explain your answer.

# pH and Buffer Solutions

## OBJECTIVES

1. To learn how to measure pH of a solution.
2. To understand the operation of buffer systems.

## BACKGROUND

We frequently encounter acids and bases in our daily life. Fruits, such as oranges, apples, etc., contain acids. Household ammonia, a cleaning agent, and Liquid Plumber® are bases. *Acids* are compounds that can donate a proton (hydrogen ion). *Bases* are compounds that can accept a proton. This classification system was proposed simultaneously by Johannes Brønsted and Thomas Lowry in 1923, and it is known as the Brønsted-Lowry theory. Thus any proton donor is an acid, and a proton acceptor is a base.

When HCl reacts with water

$$HCl + H_2O \rightleftharpoons H_3O^+ + Cl^-$$

HCl is an **acid** and $H_2O$ is a **base** because HCl **donated a proton** thereby becoming $Cl^-$, and water **accepted a proton** and thereby becoming $H_3O^+$.

In the reverse reaction (from right to left) the $H_3O^+$ is an acid and $Cl^-$ is a base. As the arrow indicates the equilibrium in this reaction lies far to the right. That is, out of every 1000 HCl molecules dissolved in water, 990 are converted to $Cl^-$ and only 10 remain in the form of HCl at equilibrium. But $H_3O^+$ (hydronium ion) is also an acid and can donate a proton to the base, $Cl^-$. Why do hydronium ions not give up protons to $Cl^-$ with equal ease and reform more HCl? This is because different acids and bases have different strengths. HCl is a stronger acid than hydronium ion, and water is a stronger base than $Cl^-$.

In the Brønsted-Lowry theory, every acid-base reaction creates its *conjugate acid-base pair*. In the above reaction HCl is an acid which, after giving up a proton, becomes a conjugate base, $Cl^-$. Similarly, water is a base which, after accepting a proton, becomes a conjugate acid, the hydronium ion.

conjugate base – acid pair

$$HCl + H_2O \rightleftharpoons H_3O^+ + Cl^-$$

conjugate acid – base pair

Some acids can give up only one proton. These are *monoprotic* acids. Examples are: **H**Cl, **H**$NO_3$, HCOO**H**, $CH_3$COO**H**. The hydrogens in **bold face** are the ones donated. Other acids yield two or three protons. These are called *diprotic* or *triprotic* acids, respectively. Examples are: $H_2SO_4$, $H_2CO_3$, and $H_3PO_4$. However, in the Brønsted-Lowry theory, each acid is considered monoprotic, and a diprotic acid (such as carbonic acid) donates its protons in two distinct steps:

$$1.\ H_2CO_3 + H_2O \rightleftharpoons H_3O^+ + HCO_3^-$$

$$2.\ HCO_3^- + H_2O \rightleftharpoons H_3O^+ + CO_3^{2-}$$

Thus the compound $HCO_3^-$ is a conjugate base in the first reaction and an acid in the second reaction. A compound that can act either as an acid or a base is called *amphiprotic*.

In the self ionization reaction

$$H_2O + H_2O \rightleftharpoons H_3O^+ + OH^-$$

one water acts as an acid (proton donor) and the other as a base (proton acceptor). In pure water, the equilibrium lies far to the left, that is, only very few hydronium and hydroxyl ions are formed. In fact, only $1\times10^{-7}$ moles of hydronium ion and $1\times10^{-7}$ moles of hydroxide ion are found in one liter of water. The dissociation constant for the self-ionization of water is

$$K_d = \frac{[H_3O^+][OH^-]}{[H_2O]^2}$$

This can be rewritten as

$$K_W = K_d[H_2O]^2 = [H_3O^+][OH^-]$$

$K_W$, the **ion product of water**, is still a constant because very few water molecules reacted to yield hydronium and hydroxide ions; hence the concentration of water essentially remained constant. At room temperature the $K_W$ has the value of

$$K_w = 1\times10^{-14} = [1\times10^{-7}]\times[1\times10^{-7}]$$

This value of the ion product of water applies not only to pure water but to any aqueous (water) solution. This is very convenient because if we know the concentration of the hydronium ion, we automatically know the concentration of the hydroxide ion and vice versa. For example, if in a 0.01 M HCl solution HCl dissociates completely, the hydronium ion concentration is $[H_3O^+] = 1\times10^{-2}\,M$. This means that the $[OH^-]$ is

$$[OH^-] = \frac{Kw}{[H_3O^+]} = \frac{1\times10^{-14}}{1\times10^{-2}} = 1\times10^{-12}\,M$$

To measure the strength of an aqueous acidic or basic solution, P.L. Sorensen introduced the pH scale.

$$pH = -\log\left[H_3O^+\right]$$

In pure water we have seen that the hydronium ion concentration is $1 \times 10^{-7}$ M. The logarithm of this is –7 and thus, the pH of pure water is 7 since [–(–7)]. Because water is an amphiprotic compound, pH 7 means a neutral solution. On the other hand, in a 0.01 M HCl solution (and dissociation is complete), we have $\left[H_3O^+\right] = 1\times10^{-2}\,M$. Thus its pH is 2. The pH scale shows that acidic solutions have a pH less than 7 and basic solutions have a pH greater than 7.

| pH | 0 1 2 3 4 5 6 7 8 9 10 11 12 13 14 |
|---|---|
| | acidic neutral basic |

The pH of an aqueous solution can be measured conveniently by special instruments called pH meters. All that must be done is to insert the electrodes of the pH meter into the solution to be measured and read the pH from a scale. pH of a solution can also be obtained, although less precisely, by using a pH indicator paper. The paper is impregnated with organic compounds that change their color at different pH values. The color shown by the paper is then compared with a color chart provided by the manufacturer.

There are certain solutions that resist a change in the pH even when we add to them acids or bases. Such systems are called *buffers*. A mixture of a weak acid and its conjugate base usually forms a good buffer system. An example is carbonic acid, which is the most important buffer in our blood and maintains it close to pH 7.4. Buffers resist large changes in pH because of the Le Chatelier principle governing equilibrium conditions. In the carbonic acid-bicarbonate (weak acid-conjugate base) buffer system,

$$H_2CO_3 + H_2O \rightleftharpoons HCO_3^- + H_3O^+$$

any addition of an acid, $H_3O^+$, will shift the equilibrium to the left. Thus this reduces the hydronium ion concentration, returning it to the initial value so that it stays constant; hence the change in pH is small. If a base, $OH^-$, is added to such a buffer system, it will react with the $H_3O^+$ of the buffer. But the equilibrium then shifts to the right, replacing the reacted hydronium ions, hence again, the change in pH is small.

Buffers stabilize a solution at a certain pH. This depends on the nature of the buffer and its concentration. For example, the carbonic acid-bicarbonate system has a pH of 6.37 when the two ingredients are at equimolar concentration. A change in the concentration of the carbonic acid relative to its conjugate base can shift the pH of the buffer. The Henderson-Hasselbalch equation below gives the relationship between pH and concentration:

$$pH = pK_a + \log\frac{\left[A^-\right]}{[HA]}$$

In this equation the $pK_a$ is the $-\log K_a$, where $K_a$ is the dissociation constant of carbonic acid

$$K_a = \frac{[HCO_3^-][H_3O^+]}{[H_2CO_3]}$$

[HA] is the concentration of the acid and [A⁻] is the concentration of the conjugate base. The $pK_a$ of the carbonic acid-bicarbonate system is 6.37. When equimolar conditions exist, then [HA] = [A⁻]. In this case the second term in the Henderson-Hasselbalch equation is zero. This is so because [A⁻]/[HA] = 1, and the log 1 = 0. Thus at equimolar concentration of the acid/conjugate base, the pH of the buffer equals the $pK_a$; in the carbonic acid-bicarbonate system this is 6.37. If, however, we have ten times more bicarbonate than carbonic acid, [A⁻]/[HA] = 10, then log 10 = 1 and the pH of the buffer will be

$$pH = pK_a + \log\frac{[A^-]}{[HA]} = 6.37 + 1.0 = 7.37$$

This is what happens in our blood—the bicarbonate concentration is ten times that of the carbonic acid and this keeps our blood at a pH of 7.4. Any large change in the pH of our blood may be fatal (acidosis or alkalosis). Other buffer systems work the same way. For example, the second buffer system in our blood is

$$H_2PO_4^- + H_2O \rightleftharpoons HPO_4^{2-} + H_3O^+$$

The $pK_a$ of this buffer system is 7.21. It requires a 1.6 to 1.0 molar ratio of $HPO_4^{2-}$ to $H_2PO_4^-$ in order to maintain our blood at pH 7.4.

## PROCEDURE

### Measurement of pH

1. Add one drop of 0.1 M HCl to the first depression of a spot plate. Dip a 2-cm long universal pH paper into the solution. Remove the excess liquid from the paper by touching the plate. Compare the color of the paper to the color chart provided (Fig. 7.1). Record the pH on your Report Sheet (1).
2. Repeat the same procedure with 0.1 M acetic acid, 0.1 M sodium acetate, 0.1 M carbonic acid (or club soda or seltzer), 0.1 M sodium bicarbonate, 0.1 M ammonia, and 0.1 M NaOH. For each solution, use a different depression of the spot plate and a new piece of pH paper. Record your results on the Report Sheet (1).

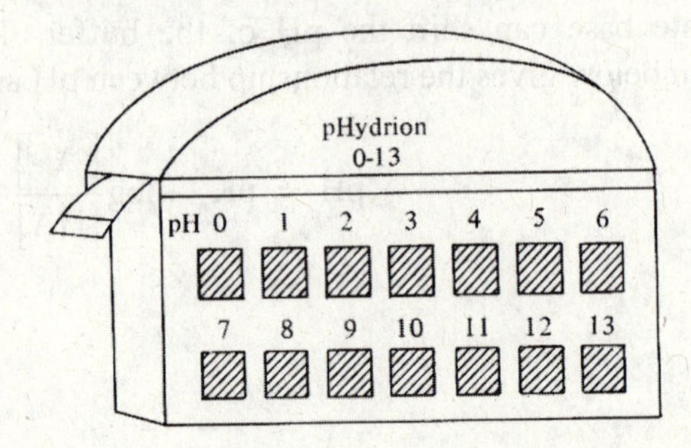


**Figure 7.1**
*pH paper dispenser.*

**CAUTION!** 

**Make sure the electrodes are immersed into the solution. Do not let the electrodes touch the walls or the bottom of the beaker. Electrodes are made of thin glass and they break easily if you do not handle them gently.**

3. Depending on the availability of the number of pH meters, this may be a class exercise (demonstration) or 6-8 students may use one pH meter. Add 5 mL of 0.1 M acetic acid to a dry and clean 10-mL beaker. Wash the electrodes over a 200-mL beaker with distilled or deionized water contained in a wash bottle. The 200-mL beaker serves to collect the wash water. Wipe gently the electrodes with Kimwipes® (or other soft tissues) to dryness. Insert the dry electrodes into the acetic acid solution. Your pH meter has been calibrated by your instructor. Switch "on" the pH meter and read the pH from the position of the needle on your scale. Alternatively, if you have a digital pH meter, a number corresponding to the pH will appear (Fig. 7.2).

4. Repeat the same procedure with 0.1 M sodium acetate, 0.1 M carbonic acid, 0.1 M sodium bicarbonate, and 0.1 M ammonia. Make certain that for each solution you use a dry and clean beaker, and before each measurement you wash the electrodes with distilled water and dry with a Kimwipe®. Record your data on the Report Sheet (2).

**Buffer Systems**

5. Prepare four buffer systems in four separate labeled dry and clean 50-mL beakers, as follows:

   a. 5 mL 0.1 M acetic acid + 5 mL 0.1 M sodium acetate

   b. 1 mL 0.1 M acetic acid + 10 mL 0.1 M sodium acetate

   c. 5 mL 0.1 M carbonic acid + 5 mL 0.1 M sodium bicarbonate

   d. 1 mL 0.1 M carbonic acid + 10 mL 0.1 M sodium bicarbonate

   Measure the pH of each buffer system with the aid of a pH meter. Record your data on the Report Sheet (3), (6), (9), and (12).

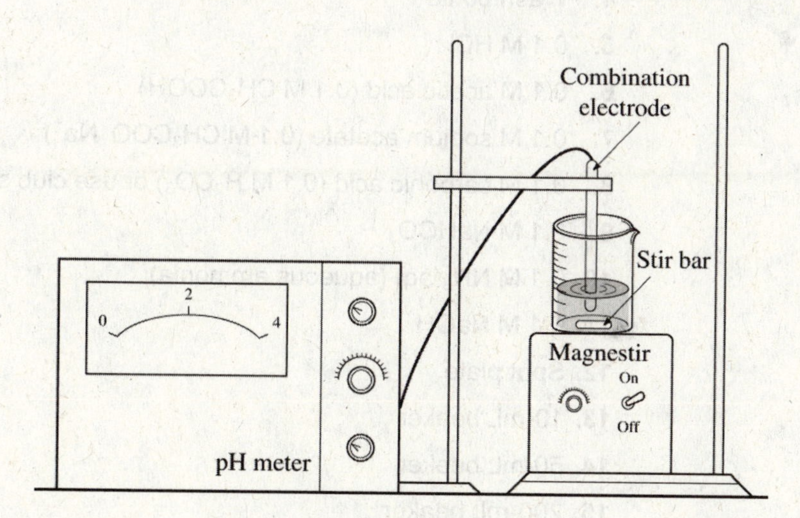


**Figure 7.2**
*pH meter.*

6. Divide each of the buffers you prepared (from above: a, b, c, d) into two halves (5 mL each) and place them into clean and dry 10-mL beakers.
   a. To the first 5-mL sample of buffer (a), add 0.5 mL 0.1 M HCl. Mix and measure the pH with the aid of a pH meter. Record your data on the Report Sheet (4).
   b. To the second 5-mL sample of buffer (a), add 0.5 mL 0.1 M NaOH. Mix and measure the pH with a pH meter. Record your data on the Report Sheet (5).
7. Repeat the same measurements, following the steps in 6. a and 6. b, using the buffers (b), (c) and (d). Be sure to use clean, dry 10-mL beakers for each preparation. Record your data on the Report Sheet for the appropriate buffer system at the spaces (7), (8), (10), (11), (13), and (14).
8. Place 5 mL of distilled water in each of two 10-mL beakers. Measure the pH of distilled water with a pH meter. Record the result on the Report Sheet (15).
   a. To the first sample of distilled water, add 0.5 mL of 0.1 M HCl. Mix and measure the pH with a pH meter. Record the result on the Report Sheet (16).
   b. To the second sample of distilled water, add 0.1 M NaOH. Mix and measure the pH as before. Record the result on the Report Sheet (17).
9. Dispose of all solutions in properly labeled liquid waste containers. Do not pour into the sink.

## CHEMICALS AND EQUIPMENT

1. pH meter
2. pHydrion® paper
3. Kimwipes®
4. Wash bottle
5. 0.1 M HCl
6. 0.1 M acetic acid (0.1 M $CH_3COOH$)
7. 0.1 M sodium acetate (0.1 M $CH_3COO^- Na^+$)
8. 0.1 M carbonic acid (0.1 M $H_2CO_3$) or use club soda or seltzer
9. 0.1 M $NaHCO_3$
10. 0.1 M $NH_3(aq)$ (aqueous ammonia)
11. 0.1 M NaOH
12. Spot plate
13. 10-mL beaker
14. 50-mL beaker
15. 200-mL beaker

name (print) | date (of lab meeting) | grade

course/section | partner's name (if applicable)

## 7 EXPERIMENT 7: pH AND BUFFER SOLUTIONS

# *Prelab Questions*

**A. Safety concerns.**

1. How does one handle the glass electrodes of a pH meter?

2. Where do you dispose of all solutions?

**B. Basic principles.**

1. Oxalic acid has the formula of HOOC–COOH (an organic acid found in rhubarb leaves as its sodium or potassium salt). It is a diprotic acid. Show the formula of the conjugate base (a) after one proton has been donated and (b) after two protons have been donated.

2. The pH of blood is 7.4 and that of saliva is 6.4. Which of the two is more acidic?

3. In general, what are the components that would make up a good buffer system?

name (print) ______ date (of lab meeting) ______ grade ______

course/section ______ partner's name (if applicable) ______

## 7 EXPERIMENT 7: pH AND BUFFER SOLUTIONS

# *Report Sheet*

| ***pH of solutions*** | **1. by pH paper** | **2. by pH meter** |
|---|---|---|
| 0.1 M HCl | ______ | not done |
| 0.1 M acetic acid | ______ | ______ |
| 0.1 M sodium acetate | ______ | ______ |
| 0.1 M carbonic acid | ______ | ______ |
| 0.1 M sodium bicarbonate | ______ | ______ |
| 0.1 M ammonia | ______ | ______ |
| 0.1 M NaOH | ______ | not done |

***Buffer Systems***

**Buffer system a** — **pH**

3. 5 mL 0.1 M $CH_3COOH$ + 5 mL 0.1 M $CH_3COO^-Na^+$ ______

4. after addition 0.5 mL 0.1 M HCl ______

5. after addition 0.5 mL 0.1 M NaOH ______

**Buffer system b**

6. 1 mL 0.1 M $CH_3COOH$ + 10 mL 0.1 M $CH_3COO^-Na^+$ ______

7. after addition 0.5 mL 0.1 M HCl ______

8 after addition 0.5 mL 0.1 M NaOH ______

**Buffer system c**

9. 5 mL 0.1 M $H_2CO_3$ + 5 mL 0.1 M $NaHCO_3$ ________________

10. after addition 0.5 mL 0.1 M HCl ________________

11. after addition 0.5 mL 0.1 M NaOH ________________

**Buffer system d**

12. 1 mL 0.1 M $H_2CO_3$ + 10 mL 0.1 M $NaHCO_3$ ________________

13. after addition 0.5 mL 0.1 M HCl ________________

14. after addition 0.5 mL 0.1 M NaOH ________________

**No buffer system**

15. distilled water ________________

16. after addition of 0.5 mL 0.1 M HCl ________________

17. after addition of 0.5 mL 0.1 M NaOH ________________

## Post-Lab Questions

1. Which of the four buffers you prepared, a, b, c, or d, is the most effective buffer system?

2. Which method gives you the best estimate of a solution's pH: by pH paper or by pH meter?

3. How does distilled water respond to the addition of acid and base? Is this a good buffer?

4. Calculate the expected pH values of the buffer systems from the experiment (a, b, c, and d), using the Henderson-Hasselbalch equation shown in the **Background**. Use the $pK_a$ values: carbonic acid/bicarbonate = 6.37 and acetic acid/acetate = 4.75.

   (a)

   (b)

   (c)

   (d)

   Are these calculated values in agreement with your measured pH values?

EXPERIMENT 8

# Structure in Organic Compounds: Use of Molecular Models

## OBJECTIVES

1. To use models to visualize structure in organic molecules.
2. To build and compare isomers having a given molecular formula.
3. To explore the three-dimensional character of organic molecules.
4. To examine a model of cyclohexane.

## BACKGROUND

The study of organic chemistry usually involves those molecules which contain carbon. Thus, a convenient definition of *organic chemistry* is the chemistry of carbon compounds.

There are several characteristics of organic compounds that make their study interesting:

a. Carbon forms strong bonds to itself as well as to other elements; the most common elements found in organic compounds, besides carbon, are hydrogen, oxygen, and nitrogen.
b. Carbon atoms are generally tetravalent. This means that carbon atoms in most organic compounds are bound by four covalent bonds to adjacent atoms.
c. Organic molecules are three-dimensional and occupy space. The covalent bonds which carbon makes to adjacent atoms are at discrete angles to each other. Depending on the type of organic compound, the angle may be 180°, 120°, or 109.5°. These angles correspond to compounds which have triple bonds, the alkynes, (1), double bonds, the alkenes, (2), and single bonds, the alkanes, (3), respectively. These compounds are called *hydrocarbons*.

$-C{\equiv}C-$ (1)　　$>C{=}C<$ (2)　　$-\overset{|}{\underset{|}{C}}-\overset{|}{\underset{|}{C}}-$ (3)

d. Organic compounds have a limitless variety in composition, shape, and structure.

Thus, while a molecular formula tells the number and type of atoms present in a compound, it tells nothing about the structure. The structural formula is a two-dimensional representation of a molecule and shows the sequence in which the atoms are connected and the bond type. For example, the molecular formula, $C_4H_{10}$, can be represented by two different structures: butane (4) and 2-methylpropane (isobutane) (5).

```
    H  H  H  H
    |  |  |  |
  H-C--C--C--C-H
    |  |  |  |
    H  H  H  H
```

Butane (4)

```
    H  H  H
    |  |  |
  H-C--C--C-H
    |  |  |
    H  |  H
     H-C-H
       |
       H
```

2-Methylpropane (5)
(Isobutane)

Consider also the molecular formula, $C_2H_6O$. There are two structures which correspond to this formula: dimethyl ether (6) and ethanol (ethyl alcohol) (7).

```
    H     H
    |     |
  H-C--O--C-H
    |     |
    H     H
```

Dimethyl ether (6)

```
    H  H
    |  |
  H-C--C-O-H
    |  |
    H  H
```

Ethanol (7)
(Ethyl alcohol)

In the pairs above, each structural formula represents a different compound. Each compound has its own unique set of physical and chemical properties. Compounds with the same molecular formula but with different structural formulas are called *isomers*.

The three-dimensional character of molecules is expressed by its stereochemistry. By looking at the *stereochemistry* of a molecule, the spatial relationships between atoms on one carbon and the atoms on an adjacent carbon can be examined. Since rotation can occur around carbon-carbon single bonds in open chain molecules, the atoms on adjacent carbons can assume different spatial relationships with respect to each other. The different arrangements that atoms can assume as a result of a rotation about a single bond are called *conformations*. A specific conformation is called a *conformer*. While individual isomers can be isolated, conformers cannot since interconversion, by rotation, is too rapid.

Conformers may be represented by the use of two conventions as shown in Fig. 8.1a and Fig 8.1b.

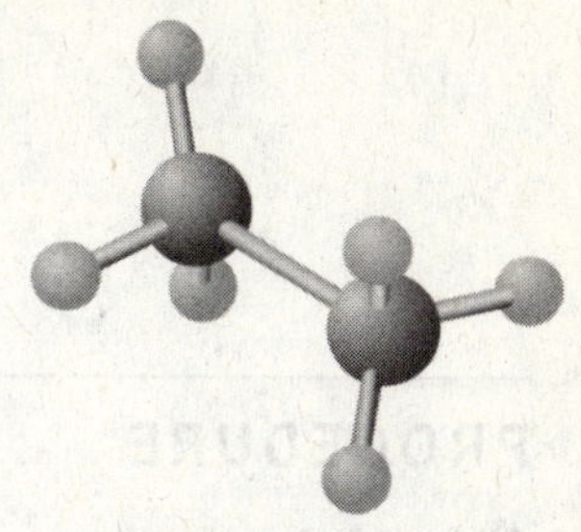

**Figure 8.1** *Molecular representations.*

**a.** Sawhorse projection of ethane. **b.** Newman projection of ethane. **c.** Ball-and-stick model of ethane.

The *sawhorse projection* views the carbon-carbon bond at an angle and, by showing all the bonds and atoms, shows their spatial arrangements. The *Newman projection* provides a view along a carbon-carbon bond by sighting directly along the carbon-carbon bond. The near carbon is represented by a circle, and bonds attached to it are represented by lines going to the center of the circle. The carbon behind is not visible (since it is blocked by the near carbon), but the bonds attached to it are partially visible and are represented by lines going to the edge of the circle. With Newman projections, rotations show the spatial relationships of atoms on adjacent carbons easily. Two conformers that represent extremes are shown in Fig. 8.2.

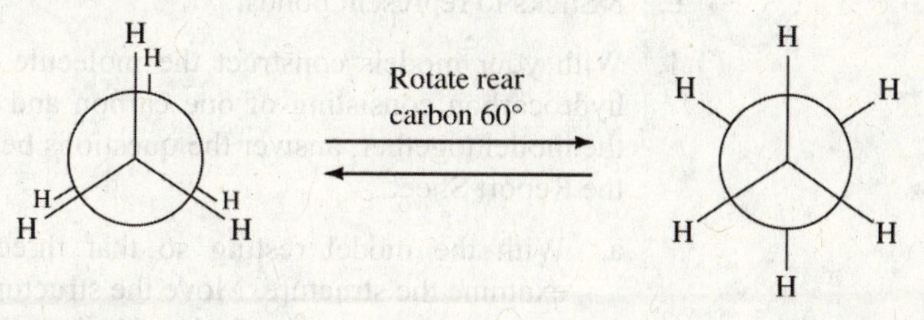


**Figure 8.2** *Two conformers of ethane.*

**a.** Eclipsed conformation of ethane. **b.** Staggered conformation of ethane.

The *eclipsed* conformation has the bonds (and the atoms) on the adjacent carbons as close as possible. The *staggered* conformation has the bonds (and the atoms) on adjacent carbons as far as possible. One conformation can interconvert into the other by rotation around the carbon-carbon bond axis.

The three-dimensional character of molecular structure is shown through molecular model building. With molecular models, the number and types of bonds between atoms and the spatial arrangements of the atoms can be visualized for the molecules. This allows comparison of isomers and of conformers for a given set of compounds. The models also will let you see what is meant by *chemical equivalence*. Here *equivalence* relates to those positions or to those hydrogens on carbon(s) in an organic molecule that are equal in terms of chemical reactivity. In the case of hydrogen, replacement of any one of the equivalent hydrogens in a molecule by a *substituent* (any atom or group of atoms, for example, Cl or OH, respectively) leads to the identical substituted molecule.

An example of a *cyclic* hydrocarbon is cyclohexane (Fig. 8.4) and is shown in what is referred to as the *chair conformation*. Here the carbon atoms are linked in a closed polygon (also referred to as a *ring*). This configuration differs from the examples discussed above, which are *chain* hydrocarbons [compounds consisting of carbons linked either in a single chain (4) or in a branched chain (5)].

## PROCEDURE

### Alkanes

Obtain a set of ball-and-stick molecular models from the laboratory instructor. The set makes alkanes and contains the following parts (other colored spheres may be substituted as available):

a. 2 Black spheres representing *Carbon*; this tetracovalent element has four holes;

b. 6 White (or Yellow) spheres representing *Hydrogen*; this monovalent element has one hole;

c. 2 Colored spheres representing the *halogen Chlorine*; this monovalent element has one hole;

d. 1 Red (or Blue) sphere representing *Oxygen*; this divalent element has two holes;

e. 8 Sticks to represent bonds.

1. With your models construct the molecule methane. Methane is a simple hydrocarbon consisting of one carbon and four hydrogens. After you put the model together, answer the questions below in the appropriate space on the Report Sheet.

   a. With the model resting so that three hydrogens are on the desk, examine the structure. Move the structure so that a different set of three hydrogens are on the desk each time. Is there any difference between the way that the two structures look (1a)?

   b. Does the term *equivalent* adequately describe the four hydrogens of methane (1b)?

   c. Tilt the model so that only two hydrogens are in contact with the desk and imagine pressing the model flat onto the desktop. Draw the way in which the methane molecule would look in two-dimensional space (1c). This is the usual way that three-dimensional structures are written.

   d. Using a protractor, measure the angle H–C–H on the model (1d).

2. Replace one of the hydrogens of the methane model with a colored sphere which represents the halogen chlorine. The new model is chloromethane (methyl chloride), $CH_3Cl$. Position the model so that the three hydrogens are on the desk.

   a. Grasp the atom representing chlorine and tilt it to the right, keeping two hydrogens on the desk. Draw the structure of the projection on the Report Sheet (2a).

b. Return the model to its original position and then tilt, as before, but this time to the left. Draw this projection on the Report Sheet (2b).

c. While the projection of the molecule changes, does the structure of chloromethane change (2c)?

3. Now replace a second hydrogen with another chlorine sphere. The new molecule is dichloromethane, $CH_2Cl_2$.

a. Examine the model as you twist and turn it in space. Are the projections given below isomers of the molecule $CH_2Cl_2$ or representations of the same structure only seen differently in three dimensions (3a)?

```
      H              Cl              H              Cl
      |              |               |              |
Cl — C — H      H — C — Cl      Cl — C — Cl     H — C — H
      |              |               |              |
      Cl             H               H              Cl
```

4. Construct the molecule ethane, $C_2H_6$. Note that you can make ethane from the methane model by removing a hydrogen and replacing the hydrogen with a methyl group, $-CH_3$.

a. Draw the structural formula for ethane (4a).

b. Are all the hydrogens attached to the carbon atoms equivalent to each other (4b)?

c. Draw a sawhorse representation of ethane. Draw a staggered and an eclipsed Newman projection of ethane (4c).

d. Replace any hydrogen in your model with chlorine. Write the structure of the molecule chloroethane (ethyl chloride), $C_2H_5Cl$ (4d).

e. Twist and turn your model. Draw two Newman projections of the chloroethane molecule (4e).

f. Do the projections that you drew represent different isomers or conformers of the same compound (4f)?

5. Dichloroethane, $C_2H_4Cl_2$

a. If you choose to remove another hydrogen from your chloroethane molecule and replace it with another chlorine, note that you now have a choice among the hydrogens. You can either remove a hydrogen from the carbon to which the chlorine is attached, or you can remove a hydrogen from the carbon that has only hydrogens attached. First, remove the hydrogen from the carbon that has the chlorine attached and replace it with a second chlorine. Draw its structure on the Report Sheet (5a).

b. Compare this structure to the model which would result from removal of a hydrogen from the other carbon and its replacement by chlorine. Draw its structure (5b) and compare it to the previous example. One isomer is 1,1-dichloroethane; the other is 1,2-dichloroethane. Label the structures drawn on the Report Sheet with the correct name.

c. Are all the hydrogens of chloroethane equivalent? Are some of the hydrogens equivalent? Label those hydrogens which are equivalent to each other (5c)?

6. Butane

a. Butane has the formula $C_4H_{10}$. With help from a partner, construct a model of butane by connecting the four carbons in a series (C–C–C–C) and then adding the needed hydrogens. First, orient the model in such a way that the carbons appear as a straight line. Next, tilt the model so that the carbons appear as a zig-zag line. Then twist around any of the C–C bonds so that a part of the chain is at an angle to the remainder. Draw each of these structures in the space on the Report Sheet (6a). Note that the structures you draw are for the same molecule but represent only a different orientation and projection.

b. Sight along the carbon-carbon bond of $\overset{2}{C}$ and $\overset{3}{C}$ on the butane chain: $\overset{1}{C}H_3-\overset{2}{C}H_2-\overset{3}{C}H_2-\overset{4}{C}H_3$ Draw a staggered Newman projection. Rotate the $C_2$ carbon clockwise by 60°; draw the eclipsed Newman projection. Again, rotate the $C_2$ carbon clockwise by 60°; draw the Newman projection. Is the last projection staggered or eclipsed (6b)? Continue rotation of the $C_2$ carbon clockwise by 60° increments and observe the changes that take place.

c. Examine the structure of butane for equivalent hydrogens. In the space on the Report Sheet (6c), redraw the structure of butane and label those hydrogens which are equivalent to each other. On the basis of this examination, predict how many monochlorobutane isomers ($C_4H_9Cl$) that could be obtained from the structure you drew in 6c (6d). Test your prediction by replacement of hydrogen by chlorine on the models. Draw the structures of these isomers (6e).

d. Reconstruct the butane system. First form a three carbon chain, then connect the fourth carbon to the center carbon of the three-carbon chain. Add the necessary hydrogens. Draw the structure of 2-methylpropane (isobutane) (6f). Can any manipulation of the model, by twisting or turning of the model or by rotation of any of the bonds, give you the butane system? If these two, butane and 2-methylpropane (isobutane), are *isomers*, then how may we recognize that any two structures are isomers (6g)?

e. Examine the structure of 2-methylpropane for equivalent hydrogens. In the space on the Report Sheet (6h), redraw the structure of 2-methylpropane and label the equivalent hydrogens. Predict how many monochloroisomers of 2-methylpropane could be formed (6i) and test your prediction by replacement of hydrogen by chlorine on the model. Draw the structures of these isomers (6j).

7. $C_2H_6O$

a. There are two isomers with the molecular formula, $C_2H_6O$, ethanol (ethyl alcohol) and dimethyl ether. With your partner, construct both of these isomers. Draw these isomers on the Report Sheet (7a) and name each one.

b. Manipulate each model. Can either be turned into the another by a simple twist or turn (7b)?

c. For each compound, label those hydrogens which are equivalent. How many sets of equivalent hydrogens are there in ethanol (ethyl alcohol) and dimethyl ether (7c)?

8. (Optional) Butenes

a. Construct 2-butene, $CH_3$–CH=CH–$CH_3$, using two springs or longer flexible grey bonds, whichever are available, for the double bond. There are two isomers for compounds of this formulation: the isomer with the 2 –$CH_3$ groups on the same side of the double bond, *cis*-2-butene; and the isomer with the 2 –$CH_3$ groups on opposite sides of the double bond, *trans*-2-butene. Draw these two structures on the Report Sheet (8a).

b. Can you twist, turn or rotate one model into the other? Explain 8b).

c. How many bonds are connected to any single carbon of these structures (8c)?

d. With the protractor measure the C–C=C angle (8d).

e. Construct methylpropene, $CH_3$–C=$CH_2$.
(with a –$CH_3$ group on the central C)

Can you have a *cis*– or a *trans*– isomer in this system (8e)?

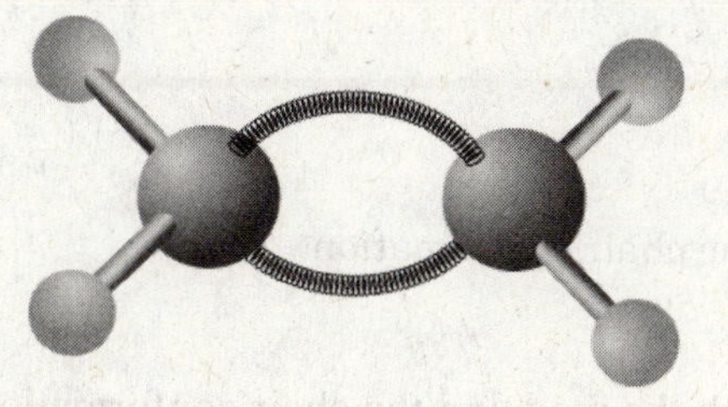

**Figure 8.3**
*Constructing a double bond.*

9. (Optional) Butynes

a. Construct 2-butyne, $CH_3$–C≡C–$CH_3$, using three springs or longer flexible grey bonds, whichever are available, for the triple bond. Can you have a *cis*- or a *trans*- isomer in this system (9a)?

b. With the protractor, measure the C–C≡C angle (9b).

c. Construct a second butyne with your molecular models and springs. How does this isomer differ from the one in (a) above (9c)?

**Cyclohexane**

For this part of the experiment, obtain a model set of "atoms" that contain the following:

a. 7 Black spheres representing *Carbon*; a model atom with 4-holes at the tetrahedral angle;

b. 1 Colored sphere representing the *halogen Chlorine*; a model atom with 1-hole;

c. 14 White (or Yellow) spheres representing *Hydrogen*; a model atom with 1-hole;

d. 21 Sticks to represent bonds.

10. Construct a model of cyclohexane by connecting 6 carbon atoms in a ring; then into each remaining hole, insert a connecting link (bond) and add a hydrogen to each.

a. Is the ring rigid or flexible, that is, can the ring of atoms move and take various arrangements in space, or is the ring of atoms locked into only one configuration (10a)?

b. Of the many configurations, which appears best for the ring: a planar or a puckered arrangement (10b)?

c. Arrange the ring atoms into a *chair* conformation (Fig. 8.4a) and compare it to the picture of a lounge chair (Fig. 8.4b). (Does the term fit the picture?)

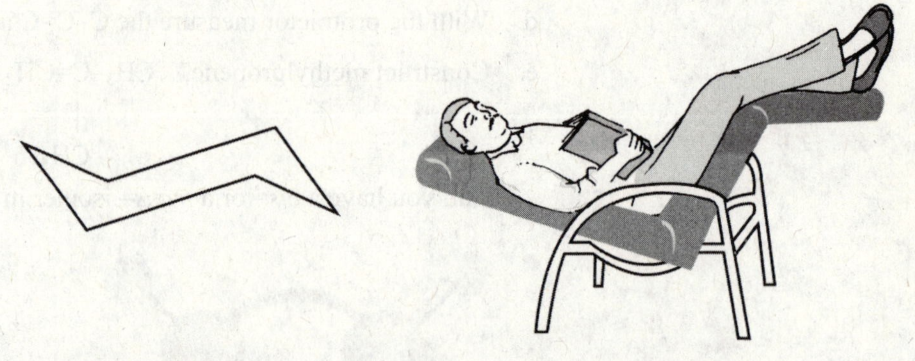

**Figure 8.4**
*The chair conformation for a 6-carbon ring.*

**a.** The chair conformation **b.** A lounge chair

11. With the model in the chair conformation, rest it on the table top.

a. How many hydrogens are in contact with the table-top (11a)?

b. How many hydrogens point in a direction 180° opposite to these (11b)?

c. Take your pencil and place it into the center of the ring perpendicular to the table. Now rotate the ring around the pencil; we'll call this an *axis of rotation*. How many hydrogens are on bonds parallel to this axis (11c)? These hydrogens are called the *axial* hydrogens and the bonds are called the *axial* bonds.

d. If you look at the perimeter of the cyclohexane system, the remaining hydrogens lie roughly in a ring perpendicular to the axis through the center of the molecule. How many hydrogens are on bonds lying in this ring (11d)? These hydrogens are called *equatorial* hydrogens and the bonds are called the *equatorial* bonds.

e. Compare your model to the diagrams in Fig. 8.5 and be sure you are able to recognize and distinguish between axial and equatorial positions.

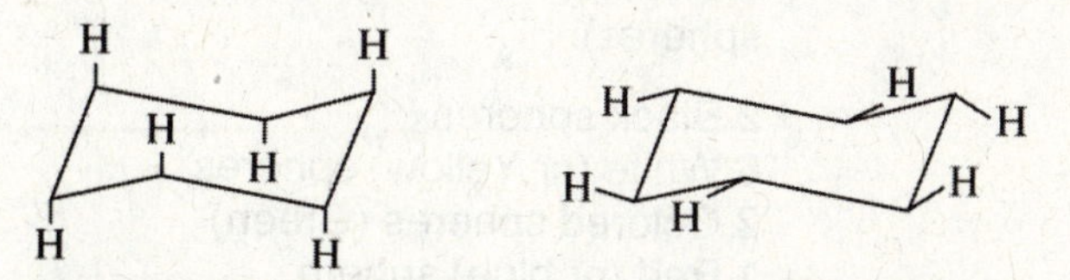


**Figure 8.5**
*Axial and equatorial hydrogens in the chair conformation.*

In the space provided on the Report Sheet (11e), draw the structure of cyclohexane in the chair conformation with all 12 hydrogens attached. Label all the axial hydrogens $H_a$ and all the equatorial hydrogens $H_e$. How many hydrogens are labeled $H_a$ (11f)? How many hydrogens are labeled $H_e$ (11g)?

12. Pick up the cyclohexane model and view it from the side of the chair. Visualize the "ring" around the perimeter of the system perpendicular to the axis through the center. Of the 12 hydrogens, how many are pointed "up" relative to the plane (12a)? How many are pointed "down" (12b)?

13. Replace one of the axial hydrogens with a colored component atom. Do a "ring flip" by moving one of the carbons *up* and moving the carbon furthest away from it *down* (Fig. 8.6). In what position is the colored component after the ring flip (13a): axial or equatorial? Do another ring flip. In what position is the colored component now (13b)? Observe all the axial positions and follow them through a ring flip.

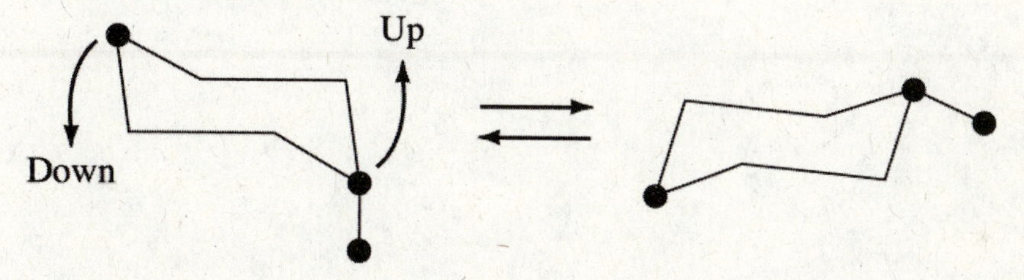


**Figure 8.6**
*A "ring flip".*

14. Replace the colored components with a methyl, $-CH_3$, group. Manipulate the model so that the $-CH_3$ group is in an axial position; examine the model. Do a ring flip placing the $-CH_3$ in an equatorial position; examine the model. Which of the chair conformations, $-CH_3$ axial or $-CH_3$ equatorial, is more crowded (14a)? What would account for one of the conformations being more crowded than the other (14b)? Which would be of higher energy, and thus, less stable (14c)? In the space provided on the Report Sheet (14d), draw the two conformations and connect with equilibrium arrows. Given your answers to 14a, 14b, and 14c, towards which conformation will the equilibrium lie (indicate by drawing one arrow bigger and thicker than the other)?

15. *A substituent group in the equatorial position of a chair conformation is more stable than the same substituent group in the axial position.* Do you agree or disagree? Explain your answer (15).

## CHEMICALS AND EQUIPMENT

### A. Alkanes

1. Molecular models (you may substitute other available colors for the spheres):

   2 Black spheres
   6 White (or Yellow) spheres
   2 Colored spheres (Green)
   1 Red (or blue) sphere
   8 Sticks

2. Protractor
3. Optional: 3 Springs or flexible grey bonds

### B. Cyclohexane

1. Molecular models (as above)

   7 Black spheres
   14 White (or Yellow) spheres
   1 Colored sphere (Green)
   21 Sticks

name (print) | date (of lab meeting) | grade

course/section | partner's name (if applicable)

8

## EXPERIMENT 8: STRUCTURE IN ORGANIC COMPOUNDS: USE OF MOLECULAR MODELS

# *Prelab Questions*

**A. Safety concerns.**

1. If this experiment is being done in the laboratory, do you still need to wear safety glasses, and should you obey all the other safety rules?

**B. Basic principles.**

1. How many bonds can each of the elements below form with neighboring atoms in a compound?

   C H O N Br S Cl

2. What is a convenient definition for organic chemistry?

3. How does a *molecular formula* differ from a *structural formula*?

4. Write structural formulas for all compounds with molecular formula $C_3H_7Cl$. How are these compounds related to each other?

5. What shape or conformation does a cyclohexane ring take?

name (print) | date (of lab meeting) | grade

course/section | partner's name (if applicable)

8

**EXPERIMENT 8: STRUCTURE IN ORGANIC COMPOUNDS: USE OF MOLECULAR MODELS**

# *Report Sheet*

**A. Alkanes.**

1. Methane

   a.

   b.

   c.

   d.

2. Chloromethane (Methyl chloride)

   a.

   b.

   c.

3. Dichloromethane

   a.

4. Ethane and chloroethane (ethyl chloride)

a.

b.

c.

d.

e.

f.

5. Dichloroethane

a.

b.

c.

6. Butane

a.

b.

c.

d.

e.

f.

g.

h.

i.

j.

7. $C_2H_6O$

a.

b.

c. Ethanol (ethyl alcohol) has___________ set(s) of equivalent hydrogens.

Dimethyl ether has__________ set(s) of equivalent hydrogens.

8. Butene

a.

b.

c.

d. C–C=C angle

e.

9. Butyne

a.

b.

c.

**B. Cyclohexane.**

10. a.

b.

11. a.

b.

c.

d.

e.

f.

g.

12. a.

b.

13. a.

b.

14 a.

b.

c.

d.

15.

## Post-Lab Questions

1. There are three (3) isomers of formula, $C_3H_8O$. Draw structural formulas for these compounds.

2. Draw the structure of propane and identify equivalent hydrogens. Identify equivalent sets by letters, e.g., $H_a$, $H_b$, etc.

3. Draw the Newman projection for one staggered and one eclipsed conformer for propane, $\overset{1}{C}H_3 - \overset{2}{C}H_2 - \overset{3}{C}H_3$, sighting along the $C_1$–$C_2$ bond.

4. Which position is more stable for the methyl group in methylcyclohexane: an equatorial position or an axial position? Explain your answer.

# Aspirin: Preparation and Properties (Acetylsalicylic Acid)

## OBJECTIVES

1. To demonstrate the synthesis of an organic compound.
2. To illustrate an isolation technique.
3. To use a chemical test to determine purity of the preparation.

## BACKGROUND

If you've ever had a headache or a muscle pain, more than likely you took an aspirin to relieve the distress. Aspirin is one of the most widely used over-the-counter, nonprescription drugs on the market. In the world, some 35,000 metric tons are produced and consumed annually, which is enough to make more than 100 billion aspirin tablets.

Aspirin was originally used as a pain killer (an analgesic) for mild headaches, toothaches, nerve pain (neuralgia), muscle pain, and joint pain due to arthritis and rheumatism. It can reduce a fever (an antipyretic) and act as an anti-inflammatory agent that can reduce the swelling and redness associated with inflammation. In addition, there are new uses for the drug. For example, aspirin is now being prescribed by many doctors to treat heart problems and to prevent strokes since it acts as an anticoagulant by preventing platelet aggregation.

Early studies showed the active agent that gave these properties was salicylic acid. However, this chemical contains a phenol group and a carboxylic acid group. Both of these groups are acidic and together are too harsh for the linings of the mouth, esophagus, and stomach. Contact with the stomach lining can cause hemorrhaging. The Bayer Company in Germany introduced and marketed a more tolerable variation of the compound. They patented the ester acetylsalicylic acid under the trademark Aspirin® in 1899. Their studies showed that this material was less of an irritant and would allow for an easier control of therapy. The acetylsalicylic acid was hydrolyzed in the small intestine to salicylic acid, which was then absorbed into the bloodstream. The relationship between salicylic acid and aspirin is shown in the following formulas:

COOH, OH (benzene ring) — Salicylic acid

COOH, O–C(=O)–$CH_3$ (benzene ring) — Acetylsalicylic acid (Aspirin)

While a century of use has confirmed its relative safety, no drug is wholly without unwanted side effects. Hemorrhaging of the stomach walls can occur even with normal dosages. Products which are better tolerated contain buffering agents, material that are bases. Magnesium hydroxide, magnesium carbonate, and aluminum glycinate, when mixed into the formulation of the aspirin (e.g., Bufferin®), reduce the irritation. Also, young children are advised not to use the drug since it can bring on a condition known as Reye's syndrome.

This experiment will acquaint you with a simple organic synthesis in the preparation of aspirin. The method transforms the phenol group into an ester by the process called acetylation, a chemical method that adds the acetyl group, $CH_3CO$. Salicylic acid reacts with acetic anhydride and an acid catalyst, like sulfuric or phosphoric acid to speed up the reaction, and is converted into acetylsalicylic acid, aspirin.

$$\text{Salicylic acid} + (CH_3CO)_2O \xrightarrow[\Delta]{H^+} \text{Aspirin} + CH_3COOH$$

Salicylic acid — Acetic anhydride — Aspirin — Acetic acid

If any salicylic acid remains unreacted, its presence can be detected with a 1% iron(III) chloride solution. Notice salicylic acid has a phenol group in its molecular structure. The iron(III) chloride gives a violet color with any molecule possessing a phenol group. Aspirin no longer has the phenol group. Thus a pure sample of aspirin will not give a purple color with 1% iron(III) chloride solution.

## PROCEDURE

### Preparation of Aspirin

1. Heat to boiling approximately 300 mL of water in a 400-mL beaker.
2. Take 2 g (measure to the nearest 0.01 g) of salicylic acid and place it in a 125-mL Erlenmeyer flask. Use this quantity to calculate the theoretical or expected yield of aspirin (1). Carefully add 3 mL of acetic anhydride to the flask and, while swirling, add 3 drops of concentrated sulfuric acid.

**CAUTION!**

**Acetic anhydride will irritate the eyes. Sulfuric acid will cause burns to the skin. Handle both chemicals with care. Use gloves with these reagents. Dispense in the hood.**

3. Mix the reagents and then place the flask into the boiling water bath; make sure that the contents are submerged in the hot water. Heat for 20 min. (Fig. 9.1). The solid will dissolve completely. Stir the solution occasionally with a glass stirring rod.

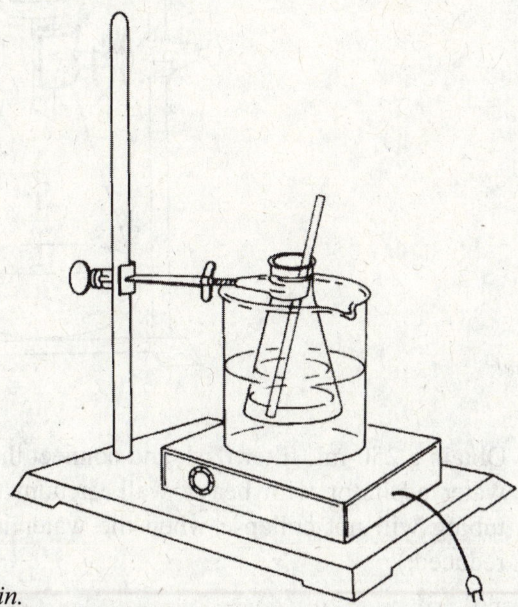

**Figure 9.1**
*Assembly for the synthesis of aspirin.*

4. Remove the Erlenmeyer flask from the bath and let it cool to approximately room temperature. Then, slowly pour the solution into a 150-mL beaker containing 20 mL of ice water, mix thoroughly, and place the beaker in an ice bath. Use a glass rod to mix the solution while in the ice bath, vigorously rubbing the glass rod (called scratching) along the bottom of the beaker (be careful not to poke a hole through the beaker). The water destroys any unreacted acetic anhydride and will cause the insoluble aspirin to precipitate from the solution. Scratching the beaker helps to precipitate the aspirin, also.
5. Collect the crystals by filtering under suction with a Büchner funnel. The assembly is shown in Fig. 9.2 and described in the following instructions.

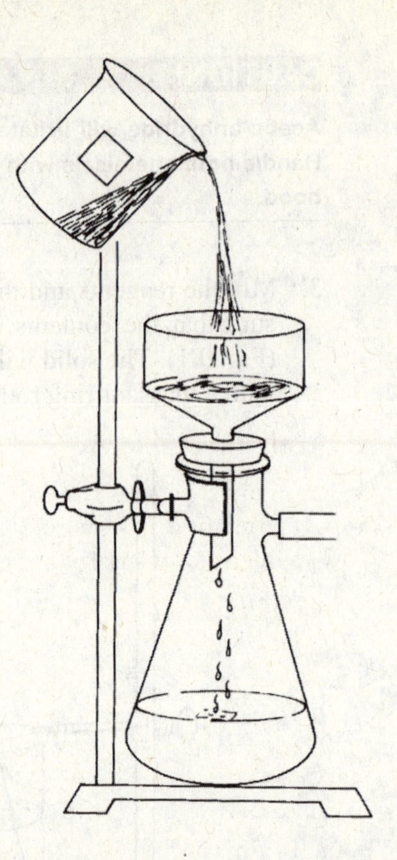

**Figure 9.2**
*Filtering using the Büchner funnel.*

6. Obtain a 250-mL filter flask and connect the side arm of the filter flask to a water aspirator with heavy wall vacuum tubing. (The thick walls of the tubing will not collapse when the water is turned on and the pressure is reduced.)

7. The Büchner funnel is inserted into the filter flask through either a filtervac, a neoprene adapter, or a one-hole rubber stopper, whichever is available. Filter paper is then placed into the Büchner funnel. Be sure that the paper lies flat and covers all the holes. Wet the paper with water.

8. Turn on the water aspirator to maximum water flow. Pour the solution into the Büchner funnel.

9. Wash the crystals with two 5 mL portions of cold water, followed by one 10 mL portion of cold ethyl acetate.

10. Continue suction through the crystals for at least 5 min. to help dry them. Disconnect the rubber tubing from the filter flask before turning off the water aspirator.

11. Using a spatula, scrape the crystals onto a sheet of paper toweling or filter paper, and press dry the solid with another piece of paper.

12. Determine the mass of a 50-mL beaker (2). Add the crystals and determine the mass again (3). Calculate the mass of crude aspirin (4). Determine the percent yield (5).

**Determine the Purity of Aspirin**

1. The aspirin you prepared is not pure enough for use as a drug and is *not* suitable for ingestion. The purity of the sample will be tested with 1% iron(III) chloride solution and compared with a commercial aspirin and salicylic acid.
2. Label three test tubes (13 × 100 mm) 1, 2, and 3; place a few crystals of salicylic acid into test tube no. 1, a small sample of your aspirin into test tube no. 2, and a small sample of crushed commercial aspirin into test tube no. 3. Add 5 mL of distilled water to each test tube and swirl to dissolve the crystals.
3. Add 10 drops of 1% aqueous iron(III) chloride to each test tube.
4. Compare and record your observations on the Report Sheet (6). The formation of a purple color indicates the presence of salicylic acid. The intensity of the color qualitatively tells how much salicylic acid is present.
5. Dispose of the material as directed by your instructor.

## CHEMICALS AND EQUIPMENT

1. Acetic anhydride
2. Concentrated sulfuric, $H_2SO_4$
3. Commercial aspirin tablets
4. Ethyl acetate
5. 1% Iron(III) chloride, aqueous solution
6. Salicylic acid
7. Boiling chips
8. Büchner funnel, small
9. 250-mL filter flask
10. Filter paper
11. Filtervac or neoprene adapter
12. Hot plate
13. Test tubes (13 × 100 mm)
14. 400-mL beaker
15. 50-mL beaker
16. 125-mL Erlenmeyer flask
17. 150-mL beaker

name (print) | date (of lab meeting) | grade

course/section | partner's name (if applicable)

## 9 EXPERIMENT 9: ASPIRIN PREPARATION AND PROPERTIES

# *Prelab Questions*

**A. Safety concerns**

1. Why dispense acetic anhydride in the hood?

2. What precaution do you take when working with sulfuric acid?

3. Would you be able to take the aspirin you prepared internally? Explain.

**B. Basic principles**

1. Salicylic acid can irritate the stomach. Name the structural features (called functional groups) that are responsible for the irritation.

2. What functional group in salicylic acid is reacted? What is the name of the new functional group that is formed in aspirin? Should the aspirin product show a color with 1% iron(III) chloride; why?

3. What is the purpose of a buffering agent in commercial preparations of aspirin?

name (print) ______ date (of lab meeting) ______ grade ______

course/section ______ partner's name (if applicable) ______

9 **EXPERIMENT 9: ASPIRIN PREPARATION AND PROPERTIES**

# *Report Sheet*

## A. Preparation of Aspirin

1. Theoretical yield:

$$\_\_\_\_\_\_ \text{ g salicylic acid} \times \frac{1 \text{ mole}}{138 \text{ g salicylic acid}} \times \frac{180 \text{ g aspirin}}{1 \text{ mole}} = \_\_\_\_\_\_ \text{ g aspirin}$$

2. Mass of 50-mL beaker ______ g

3. Mass of your aspirin and beaker ______ g

4. Mass of your aspirin: (3) – (2) ______ g

5. Percent yield: $\frac{(4)}{(1)} \times 100 = \%$ ______ %

## B. Determine the Purity of the Aspirin

| *No.* | *Sample* | *Color* | *Intensity* |
|---|---|---|---|
| **1** | Salicylic acid | | |
| **2** | Your aspirin | | |
| **3** | Commercial aspirin | | |

## Post-Lab Questions

1. The standard dose of aspirin in a tablet is 5 grains. How many mg is this? (1 grain = 0.0648 g)

2. Some nonprescription pain relievers available as substitutes for aspirin are acetaminophen (sold as Tylenol®), ibuprofen (sold as Advil®, Motrin®), and naproxen (sold as Bonyl®, Prexan®). Which drug would give a positive test for phenol? How do you know?

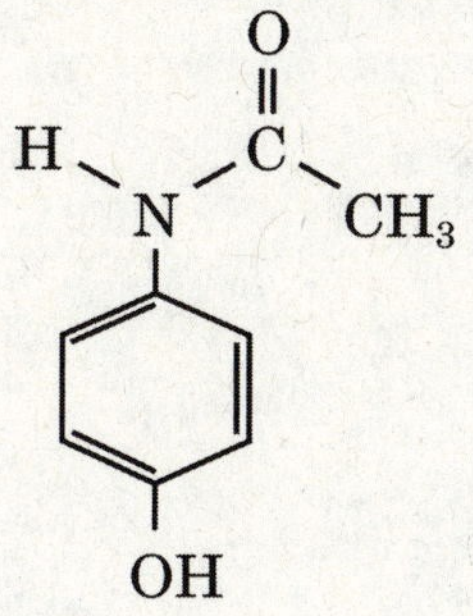

Acetaminophen 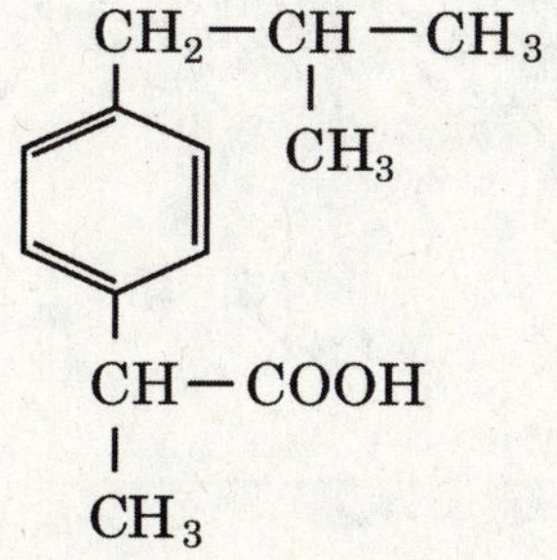 | Ibuprofen 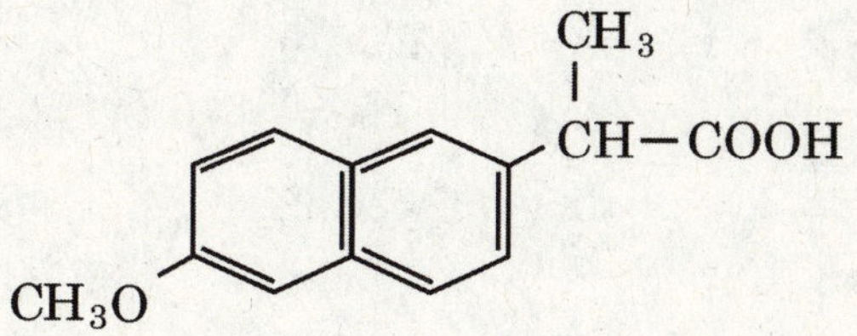 | Naproxen

3. A student expected a theoretical yield of 2.53 g of acetylsalicylic acid, but isolated 3.05 g. What could account for the 'extra' material? What step may have been omitted?

4. Aspirin tablets from an opened, old bottle smelled of vinegar and gave a purple color with 1% iron(III) chloride solution. What chemical gave that odor? Which functional group in the aspirin reacted to form the chemical that smells of vinegar?

5. If a student doubled the amount of acid used in the experiment, will the yield of acetylsalicylic acid change? Explain.

EXPERIMENT 10

# Carbohydrates

## OBJECTIVES

1. To investigate some chemical properties of carbohydrates in terms of their functional groups.
2. To become familiar with the reducing or non-reducing nature of carbohydrates.
3. To compare monosaccharides, disaccharides, and polysaccharides.
4. To experience the acid-catalyzed hydrolysis of acetal groups and the relation to digestion.

## BACKGROUND

Carbohydrates are a major food source. Rice, potatoes, bread, corn, candy, and fruits are rich in carbohydrates. A carbohydrate can be classified as a monosaccharide (for example glucose or fructose), a disaccharide (sucrose or lactose), which consists of two joined monosaccharides, or a polysaccharide (starch or cellulose), which consists of thousands of monosaccharide units linked together. If you look at the functional groups present, carbohydrates are polyhydroxy aldehydes or ketones or compounds that yield polyhydroxy aldehydes or ketones upon hydrolysis. Monosaccharides exist mostly as cyclic structures containing hemiacetal (or hemiketal) groups. These structures in solutions are in equilibrium with the corresponding open chain structures bearing aldehyde or ketone groups. Glucose, blood sugar, is an example of a polyhydroxy aldehyde (Fig. 10.1).

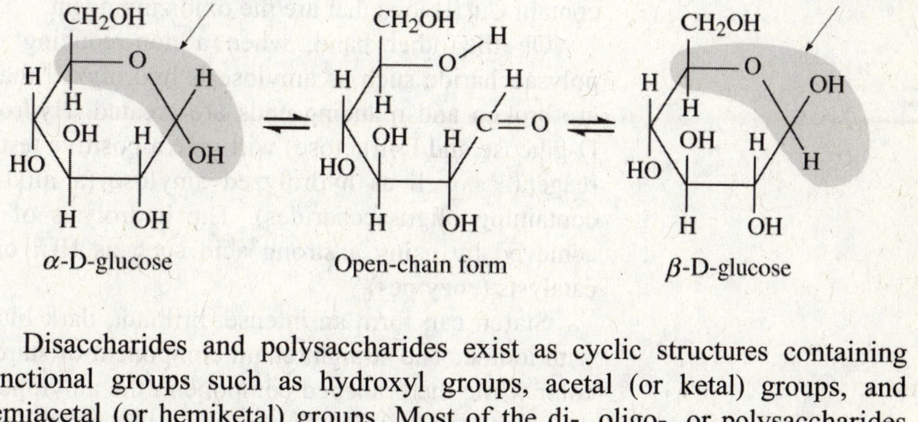

**Figure 10.1**
*The structures of D-glucose.*

Disaccharides and polysaccharides exist as cyclic structures containing functional groups such as hydroxyl groups, acetal (or ketal) groups, and hemiacetal (or hemiketal) groups. Most of the di-, oligo-, or polysaccharides have two distinct ends. The one end which has a hemiacetal (or hemiketal) on its terminal is called the reducing end, and the one which does not contain a hemiacetal (or hemiketal) terminal is the non-reducing end. The name "reducing" is given because hemiacetals (and to a lesser extent hemiketals) can reduce an oxidizing agent such as Fehling's reagent (and has an “unlocked” ring).

Fig. 10.2 is an example of a disaccharide with a hemiacetal and an acetal in it:

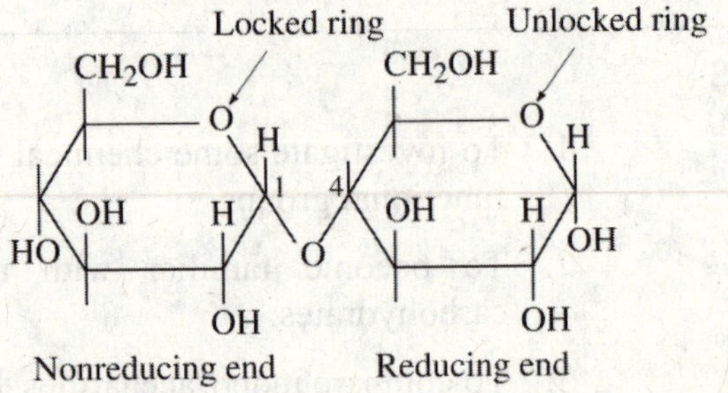

**Figure 10.2**
*The structure of maltose, a disaccharide.*

Not all disaccharides or polysaccharides contain a reducing end. An example is sucrose, which does not have a hemiacetal (or hemiketal) group on either of its ends (Fig. 10.3).

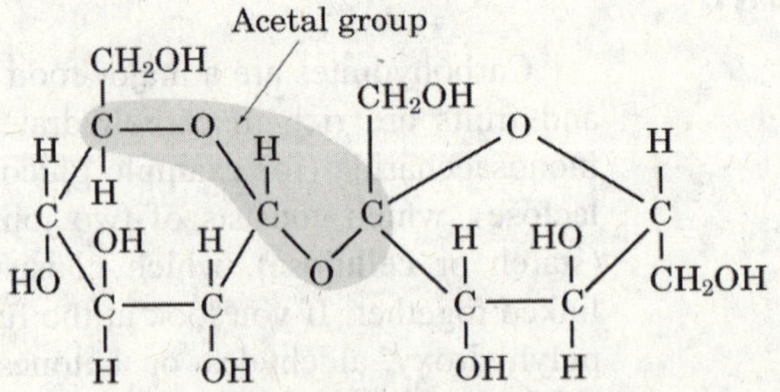

**Figure 10.3**
*The structure of sucrose.*

Polysaccharides, such as amylose or amylopectin, do have a hemiacetal group on one of their terminal ends, but practically, they are non-reducing substances because there is only one reducing group for every 2,000-10,000 monosaccharidic units. In such a low concentration, the reducing group does not give a positive test with Benedict's or Fehling's reagent; these reagents contain Cu(II) ions that are the oxidizing agent.

On the other hand, when a non-reducing disaccharide (sucrose) or a polysaccharide such as amylose is hydrolyzed the glycosidic linkages (acetal) are broken and reducing ends are created. Hydrolyzed sucrose (a mixture of D-glucose and D-fructose) will give a positive test with Benedict's or Fehling's reagent as well as hydrolyzed amylose (a mixture of glucose and glucose containing oligosaccharides). The hydrolysis of sucrose or amylose can be achieved by using a strong acid such as HCl or with the aid of biological catalysts (enzymes).

Starch can form an intense, brilliant, dark blue or violet colored complex with iodine. The straight chain component of starch, the amylose, gives a blue color while the branched component, the amylopectin yields a purple color. In the presence of iodine the amylose forms helixes inside of which the iodine molecule assemble as long polyiodide chains. The helix forming branches of amylopectin are much shorter than those of amylose. Therefore, the polyiodide chains are also much shorter in the amylopectin-iodine complex than in the amylose-iodine complex. The result is a different color (purple). When starch is hydrolyzed and broken down to small carbohydrate units the iodine will not give a dark blue (or purple) color. The iodine test is used in this experiment to indicate the completion of the hydrolysis.

In this experiment you will investigate some chemical properties of carbohydrates in terms of their functional groups.

1. Reducing and non-reducing properties of carbohydrates

**a. Aldoses (polyhydroxy aldehydes).** All aldoses are reducing sugars because they contain free aldehyde functional groups. The aldehydes are oxidized by mild oxidizing agents (e.g., Benedict's or Fehling's reagent) to the corresponding carboxylates. For example,

$$\underset{\text{(from Fehling's reagent)}}{\text{R—CHO} + 2\text{Cu}^{2+}} \xrightarrow{\text{NaOH}} \text{R—COO}^-\text{Na}^+ + \underset{\text{Red precipitate}}{\text{Cu}_2\text{O}\downarrow}$$

**b. Ketoses (polyhydroxy ketones).** All ketoses are reducing sugars because they have a ketone functional group next to an alcohol functional group. The reactivity of this specific ketone (also called an α-hydroxyketone) is attributed to its ability to form an α-hydroxyaldehyde in basic media according to the following equilibrium equations:

$$\underset{\text{Ketose}}{\begin{array}{c}\text{CH}_2\text{OH}\\ |\\ \text{C=O}\\ |\\ \text{H—C—OH}\\ |\end{array}} \underset{}{\overset{\text{base}}{\rightleftharpoons}} \underset{\text{Enediol}}{\begin{array}{c}\text{CHOH}\\ \|\\ \text{C—OH}\\ |\\ \text{H—C—OH}\\ |\end{array}} \overset{\text{base}}{\rightleftharpoons} \underset{\text{Aldose}}{\begin{array}{c}\text{CHO}\\ |\\ \text{H—C—OH}\\ |\\ \text{H—C—OH}\\ |\end{array}}$$

**c. Hemiacetal functional group (potential aldehydes).** Carbohydrates with hemiacetal functional groups can reduce mild oxidizing agents such as Fehling's reagent because hemiacetals can easily form aldehydes through the following equiliblium equation:

$$\begin{array}{c}\text{H}\quad\text{OR}'\\ \diagdown\ \diagup\\ \text{C}\\ \diagup\ \diagdown\\ \text{R}\quad\text{OH}\end{array} \rightleftharpoons \begin{array}{c}\text{O}\\ \|\\ \text{R—C—H}\end{array} + \text{R}'\text{OH}$$

Sucrose, on the other hand, is a non-reducing sugar because it does not contain a hemiacetal functional group. Although starch has a hemiacetal functional group at one end of its molecule, it is, however, considered as a non-reducing sugar because the effect of the hemiacetal group in a very large starch molecule becomes too insignificant to give a positive Fehling's (or Benedict's) test.

2. Hydrolysis of acetal groups. Disaccharides and polysaccharides can be converted into monosaccharides by hydrolysis. The following is an example:

$$\underset{\substack{\text{Lactose}\\\text{(milk sugar)}}}{C_{12}H_{22}O_{11}} + H_2O \xrightarrow{\text{catalyst}} \underset{\text{Glucose}}{C_6H_{12}O_6} + \underset{\text{Galactose}}{C_6H_{12}O_6}$$

## PROCEDURE

### Part A. Reducing or Non-Reducing Carbohydrates

1. Place approximately 2 mL (approximately 40 drops) of Fehling's solution (20 drops each of solution part A and solution part B) into each of five labeled tubes.
2. Add 10 drops of each of the following carbohydrates to the corresponding test tubes as shown in the following table.

| *Test Tube No.* | *Name of Carbohydrate* |
|---|---|
| 1 | Glucose |
| 2 | Fructose |
| 3 | Sucrose |
| 4 | Lactose |
| 5 | Starch |

3. Place the test tubes in a boiling water bath for 5 min. A 600-mL beaker containing about 200 mL of tap water and a few boiling chips is used as the bath. Record your results on your Report Sheet. Which of those carbohydrates are reducing carbohydrates?

**CAUTION!**

**Remember to use the boiling chips; they prevent bumping. Handle the hot test tubes with a test tube holder and the hot beaker with beaker tongs.**

### Part B. Hydrolysis of Carbohydrates

Hydrolysis of sucrose (acid versus base)

1. Place 3 mL of 2% sucrose solution in each of two labeled test tubes. To the first test tube (no. 1), add 3 mL of water and 3 drops of dilute sulfuric acid solution (3 M $H_2SO_4$). To the second test tube (no. 2), add 3 mL of water and 3 drops of dilute sodium hydroxide solution (3 M NaOH).

**CAUTION!**

**To avoid burns from the acid or the base, use gloves when dispensing these reagents.**

2. Heat the test tubes in a boiling water bath for about 5 min. Then cool both solutions to room temperature.
3. To the contents of test tube no. 1, add dilute sodium hydroxide solution (3 M NaOH) (about ten drops) until red litmus paper turns blue.
4. Test a few drops of each of the two solutions (test tubes nos. 1 and 2) with Fehling's reagent following the procedure that is described for carbohydrates in Part A (above). Record your results on your Report Sheet.

<u>Acid catalyzed hydrolysis of starch</u>

1. Place 5.0 mL of starch solution in a 15 × 150 mm test tube and add 1.0 mL of dilute sulfuric acid (3 M $H_2SO_4$). Mix it by gently shaking the test tube. Heat the solution in a boiling water bath for about 5 min.
2. Using a clean medicine dropper, transfer about 3 drops of the starch solution into a white spot plate and then add 2 drops of iodine solution. Observe the color of the solution. If the solution gives a positive test with iodine solution (the solution should turn blue), the hydrolysis is not complete and you should continue heating.
3. Transfer about 3 drops of the boiling solution at 5-min. intervals for an iodine test. <u>(Note: Rinse the medicine dropper very thoroughly before each test</u>.) When the solution no longer gives the characteristic blue color with iodine solution, stop heating and record the time needed for the completion of hydrolysis on the Report Sheet.

## CHEMICALS AND EQUIPMENT

1. Bunsen burner (or hot plate)
2. Medicine droppers
3. White spot plate
4. Boiling chips
5. Fehling's reagent
6. 3 M NaOH
7. 2% starch solution
8. 2% sucrose
9. 2% fructose
10. 2% glucose
11. 2% lactose
12. 3 M $H_2SO_4$
13. 0.01 M iodine in KI
14. Test tube (15 × 150 mm)
15. 600-mL beaker

name (print) | date (of lab meeting) | grade

course/section | partner's name (if applicable)

10 **EXPERIMENT 10: CARBOHYDRATES**

# *Prelab Questions*

## A. Safety concerns

1. Why do you use boiling chips when you are boiling a liquid?

2. How should you handle the hot glassware in this experiment?

## B. Basic principles

1. What functional group(s) are present in reducing carbohydrates?

2. Circle and label the hemiacetal functional group and the acetal functional group in each of the following carbohydrates:

**a.** sucrose

**b.** lactose

3. Which carbohydrate in question 2 is a reducing sugar?

4. Name the functional group that links two monosaccharides in a disaccharide.

name (print) | date (of lab meeting) | grade

course/section | partner's name (if applicable)



# Report Sheet

## Reducing or Non-Reducing Carbohydrates

| *Test Tube No.* | *Substance* | *Color Observation* | *Reducing or Non-Reducing Carbohydrate* |
|---|---|---|---|
| 1 | Glucose | | |
| 2 | Fructose | | |
| 3 | Sucrose | | |
| 4 | Lactose | | |
| 5 | Starch | | |

## Hydrolysis of Carbohydrates

### Hydrolysis of Sucrose (Acid Versus Base Catalysis)

| *Sample* | *Condition of Hydrolysis* | *Color Observation* | *Fehling's Test (Positive or Negative)* |
|---|---|---|---|
| 1 | Acidic ($H_2SO_4$) | | |
| 2 | Basic (NaOH) | | |

### Acid Catalyzed Hydrolysis of Starch

| *Sample No.* | *Heating Time (min)* | *Color Observation* | *Iodine Test (Positive or Negative)* |
|---|---|---|---|
| 1 | 5 | | |
| 2 | 10 | | |
| 3 | 15 | | |
| 4 | 20 | | |
| 5 | 25 | | |
| 6 | 30 | | |

## Post-Lab Questions

1. How does the iodine test distinguish between amylose and amylopectin?

2. Why is sucrose a non-reducing sugar?

3. How can you tell when the hydrolysis of starch is complete? Why does the test work this way? What is the monosaccharide that results at the end?

4. Why does amylose give a negative test with Fehling's solution?

# *Fats and Oils: Preparation and Properties of a Soap*

## OBJECTIVES

1. To study the properties of triglycerides.
2. To prepare a simple soap.
3. To investigate some properties of a soap.

## BACKGROUND

Fats and oils are triglycerides, esters of glycerol and fatty acids. Fats have large proportions of long chain saturated fatty acids; an example is stearic acid. Oils, on the other hand, contain mostly long chain unsaturated fatty acids; an example is oleic acid. Some oils like coconut oil contain lots of short chain saturated fatty acids; an example is lauric acid.

| | |
|---|---|
| $CH_3(CH_2)_{16}COOH$ | Stearic acid |
| $CH_3(CH_2)_7CH{=}CH(CH_2)_7COOH$ | Oleic acid |
| $CH_3(CH_2)_{10}COOH$ | Lauric acid |

**Figure 11.1**
*Some saturated and unsaturated fatty acids.*

Saturated fatty acids and triglycerides containing saturated fatty acids are solids at room temperature because the regular nature of their aliphatic chains allows the molecules to be packed in a close, parallel alignment. The longer the aliphatic chain, the higher the melting point. The interactions between neighboring chains are weak London dispersion forces. Nevertheless, the regular packing allows these forces to operate over a large portion of the chain, ensuring that a considerable amount of energy is needed to melt them.

$$\begin{array}{ll} CH_2{-}O{-}\overset{\displaystyle O}{\overset{\|}{C}}{-}(CH_2)_{16}{-}CH_3 & \text{stearic acid} \\ | & \\ CH{-}O{-}\overset{\displaystyle O}{\overset{\|}{C}}{-}(CH_2)_2{-}CH{=}CH{-}(CH_2)_7{-}CH_3 & \text{oleic acid} \\ | & \\ CH{-}O{-}\underset{\displaystyle O}{\underset{\|}{C}}{-}(CH_2)_{10}{-}CH_3 & \text{lauric acid} \end{array}$$

**Figure 11.2**
*A triglyceride containing 3 different* fatty acids.

Unsaturated fatty acids and oils containing them, in contrast are all liquids at room temperature because the cis double bonds interrupt the regular packing of the chains.

Nutritionists recommend a diet rich in unsaturated fatty acids, especially omega fatty acids: omega-3 or omega-6 fatty acids. The omega refers to the last carbon of the hydrocarbon chain. Omega-3 fatty acids have their "first" double bonds on the third carbon atom counting from the omega end; an example is linolenic acid. An omega-6 fatty acid, such as linoleic acid, has its "first" double bond on the 6th carbon counting from the tail end.

$\omega$–3
$CH_3CH_2CH{=}CHCH_2(CH{=}CHCH_2)_2(CH_2)_6COOH$ Linolenic acid

(Omega end)

$\omega$–6
$CH_3(CH_2)_4CH{=}CHCH_2CH{=}CHCH_2(CH_2)_6COOH$ Linoleic acid

An easy way to test for unsaturated fatty acids is the bromine test which measures the degree of unsaturation by reacting with the double bonds of the fatty acids (equation 1).

$$CH_3(CH_2)_7CH{=}CH(CH_2)_7COOH + \underset{\text{Red}}{Br_2} \rightarrow \underset{\text{Colorless}}{CH_3(CH_2)_7\underset{Br}{CH}{-}\underset{Br}{CH}(CH_2)_7COOH} \quad (1)$$

Quantitatively, the amount of $Br_2$ added to secure a permanent red color is a measure of the number of –C=C– reacted. If there are only saturated fatty acids present, the red color will persist after the addition of a few drops of bromine. The more unsaturated fatty acids that are in the structure of an oil, the more bromine required to produce a stable red color.

A soap is the sodium or potassium salt of a long-chain fatty acid. The fatty acid usually contains 12 to 18 carbon atoms. Solid soaps usually consist of sodium salts of fatty acids, whereas liquid soaps consist of the potassium salts of fatty acids. A soap, such as sodium stearate, consists of a nonpolar end (the hydrocarbon chain of the fatty acid) and a polar end (the ionic carboxylate).

$$CH_3CH_2CH_2CH_2CH_2CH_2CH_2CH_2CH_2CH_2CH_2CH_2CH_2CH_2CH_2CH_2CH_2{-}\overset{O}{\overset{\|}{C}}{-}O^-Na^+$$

Nonpolar; hydrophobic (dissolves in oils) — Polar; hydrophilic (dissolves in water)

**Figure 11.3**
*A soap (sodium stearate).*

Because "like dissolves like," the nonpolar end (hydrophobic or water hating part) of the soap molecule can dissolve the greasy dirt, and the polar or ionic end (hydrophilic or water loving part) of the molecule is attracted to water molecules. Therefore, the dirt from the surface being cleaned will be pulled away and suspended in water. Thus, soap acts as an emulsifying agent, a substance used to disperse one liquid (oil molecules) in the form of finely suspended particles or droplets in another liquid (water molecules).

Treatment of fats or oils with strong bases such as NaOH or KOH causes them to undergo hydrolysis (saponification) to form glycerol and the salts of a long-chain fatty acid (soap) (equation 2).

$$\begin{array}{l} CH_2-O-\overset{\overset{O}{\parallel}}{C}-C_{17}H_{35} \\ | \\ CH-O-\overset{\overset{O}{\parallel}}{C}-C_{17}H_{35} \\ | \\ CH_2-O-\overset{\overset{O}{\parallel}}{C}-C_{17}H_{35} \end{array} + 3NaOH \xrightarrow{\Delta} \begin{array}{c} CH_2OH \\ | \\ CHOH \\ | \\ CH_2OH \end{array} + 3C_{17}H_{35}\overset{\overset{O}{\parallel}}{C}-O^-Na^+ \quad (2)$$

Tristearin — Glycerol — Sodium stearate (a soap)

Because soaps are salts of strong bases and weak acids, they should be weakly alkaline in aqueous solution. However, a soap with free alkali can cause damage to skin, silk, or wool. Therefore, a test for basicity is quite important.

$$CH_3(CH_2)_{11}-O-\underset{\underset{O}{\parallel}}{\overset{\overset{O}{\parallel}}{S}}-O^-\ Na^+ \quad (3)$$

A synthetic detergent

Soap has been largely replaced by synthetic detergents (equation 3; sodium lauryl sulfate, an organic detergent) during the last two decades because soap has serious drawbacks. One drawback is that soap becomes ineffective in hard water because hard water contains appreciable amounts of $Ca^{2+}$ or $Mg^{2+}$ cations. The calcium or magnesium salts are insoluble in water and precipitate (↓) as scum (so called 'ring around the tub') (equation 4).

$$\underset{\text{Soap}}{2C_{17}H_{35}COO^-Na^+} + M^{2+} \longrightarrow \underset{\text{Scum}}{[C_{17}H_{35}COO^-]_2\, M^{2+}\downarrow} + 2Na^+ \quad (4)$$

$$M = (Ca^{2+} \text{ or } Mg^{2+})$$

The second drawback is that, in an acidic solution, soap is converted to free fatty acid and therefore, loses its charge, its water solubility, and cleansing action (equation 5). Detergents are not affected by either of these problems.

$$\underset{\text{Soap}}{C_{17}H_{35}COO^{-}Na^{+}} + H^{+} \longrightarrow \underset{\text{Fatty acid}}{C_{17}H_{35}COOH\downarrow} + Na^{+} \quad (5)$$

## PROCEDURE

### Testing for Unsaturated Fatty Acids

1. Label 3 test tubes (13 × 100 mm) no. 1, no. 2, and no. 3. Add 5 drops of corn oil to no.1, a similar quantity of butter to no. 2, and 5 drops of mineral oil (which is a mixture of long chain alkanes) to no. 3. Add 1 mL (about 20 drops) of dichloromethane to each test tube.

2. Shake and swirl the test tubes. Note whether the substances are soluble or not in the solvent. If not soluble, add another 0.5 mL of dichloromethane to each of the test tubes. Record the solubility on the Report Sheet (1).

3. To test tube no. 1, add dropwise and with shaking, 1% $Br_2$ in cyclohexane. Count the number of drops. Note if the red color persists after each addition. If the color fades away, add more drops until the red color becomes permanent. Record the number of drops of bromine solution added on your Report Sheet (2).

4. Repeat the same procedure (as in step no. 3 above) with test tube no. 2 and test tube no. 3. Record the results on your Report Sheet in (3) and (4).

**CAUTION!**

**Use the 1% $Br_2$ solution in the hood. Wear gloves when using this chemical. Bromine is poisonous. Do not breathe the fumes. Avoid contact; it can cause burns.**

### Preparation of a Soap

1. Measure 23 mL of a vegetable oil into a 250-mL Erlenmeyer flask. Add 20 mL of ethanol (to act as a solvent) and 20 mL of 25 % NaOH solution. While stirring the mixture constantly with a glass rod, the flask with its contents is heated gently in a boiling water bath. A 600-mL beaker containing about 200 mL of tap water and a few boiling chips can serve as a water bath (Fig. 11.4).

**CAUTION!**

**Ethanol is flammable. No open flames should be in the laboratory. The 25% NaOH solution is caustic and can cause severe burns if it gets into the eyes. Wear goggles all the time. Wash the base off immediately if any gets on your skin.**

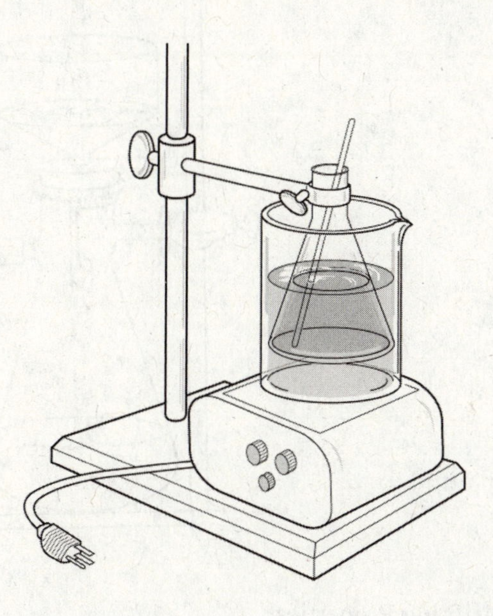

**Figure 11.4**
*Experimental set-up for the soap preparation.*

2. After heating for 20 min., the odor of the ethanol will disappear, indicating the completion of the reaction. A pasty mass containing a mixture of the soap, glycerol, and excess sodium hydroxide is obtained.
3. Cool the flask with its contents in an ice-water bath to room temperature.
4. Pour the mixture into a 400-mL beaker containing 150 mL of a saturated sodium chloride solution while stirring vigorously. This precipitates or "salts out" the soap; the process increases the density of the aqueous solution so that the soap floats out.
5. Filter the precipitated soap with the aid of a suction filtration set-up and a Büchner funnel (Fig. 11.5). Wash the precipitate with 10 mL of ice cold water.
6. Observe the appearance of your soap and record your observation on the Report Sheet (5).

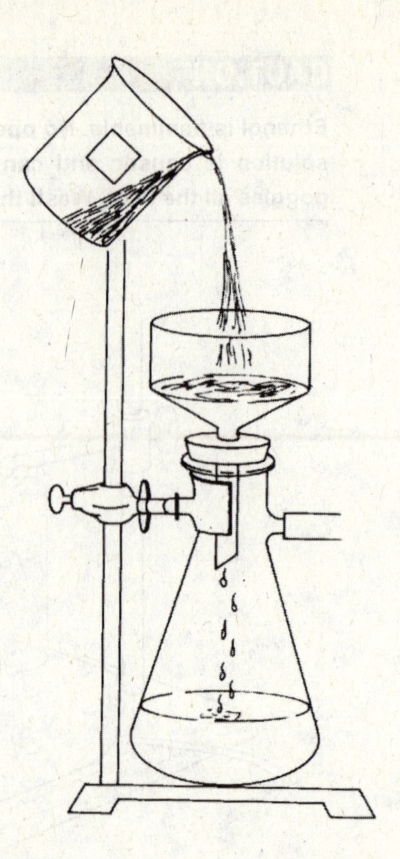

**Figure 11.5**
*Vacuum filtration with a Büchner funnel.*

**Properties of a Soap**

**(a)** Emulsifying Properties.

1. Shake 5 drops of mineral oil in a test tube (13 × 100 mm) containing 5 mL of water. A temporary emulsion of tiny oil droplets in water will be formed.
2. Repeat the same test, but this time, add a small piece of the soap you have prepared before you shake. Allow both solutions to stand for awhile.
3. Compare the appearance and the relative stabilities of the two emulsions. Record your observations on the Report Sheet (6).

**(b)** Hard Water Reactions.

4. Place a small amount (an amount on one-third of the tip of a small spatula) of the soap you prepared in a 50-mL beaker containing 25 mL of distilled (or deionized) water. Warm the beaker with its contents to dissolve the soap.
5. Pour 5 mL of the diluted soap solution into each of four test tubes (13 × 100 mm): nos. 1, 2, 3, and 4.
6. To test tube no. 1, add 2 drops of 5% $CaCl_2$ solution; to test tube no. 2, add 2 drops of 5% $MgCl_2$ solution; to test tube no. 3, add 1 mL of tap water. The solution in test tube no. 4 will be used for a test of basicity to be performed later.
7. Record your observations on the Report Sheet (7), (8), and (9).

**(c)** Alkalinity (Basicity).

8. Test the soap solution in test tube no. 4 with a wide-range pH paper. What is the approximate pH of your soap solution? Record your answer on the Report Sheet (10).

## CHEMICALS AND EQUIPMENT

1. Boiling chips
2. Büchner funnel in a No. 7 one-hole rubber stopper
3. Crushed ice
4. 500-mL filter flask
5. Filter paper, Whatman no.2, 7 cm diameter
6. Hot plate
7. Wide-range pH paper
8. Butter
9. 1% $Br_2$ in cyclohexane (bromine)
10. 5% $CaCl_2$ solution (calcium chloride)
11. Corn oil
12. Dichloromethane, $CH_2Cl_2$
13. 95% ethanol, $CH_3CH_2OH$
14. Mineral oil
15. 5% $MgCl_2$ solution (magnesium chloride)
16. Saturated sodium chloride solution, NaCl
17. 25% NaOH (sodium hydroxide)
18. Test tubes (13 × 100 mm)
19. 250-mL Erlenmeyer flask
20. 600-mL beaker
21. 400-mL beaker
22. 50-mL beaker

name (print) date (of lab meeting) grade

course/section partner's name (if applicable)

11 **EXPERIMENT 11: FATS AND OILS**

# *Prelab Questions*

**A. Safety concerns.**

1. Why must you perform the bromine addition under the hood?

2. How can you avoid contact with bromine solution?

3. What precautions do we have to take when using ethanol?

4. What protection do you need when using 25% NaOH solution?

**B. Basic principles.**

1. Define the following terms:

   a. Hydrophobic

   b. Hydrophilic

2. How many moles of $Br_2$ are needed to saturate 1 mole of oleic acid?

3. How would you convert a potassium oleate soap to the corresponding fatty acid?

name (print) ______ date (of lab meeting) ______ grade ______

course/section ______ partner's name (if applicable) ______

11 **EXPERIMENT 11: FATS AND OILS**

# *Report Sheet*

**A. Bromine test.**

Volume in mL of dichloromethane required to dissolve:

1. corn oil__________; butter__________; mineral oil__________.

Number of drops of 1% $Br_2$ solution required to produce a permanent red color with

2. corn oil__________; 3. butter__________; 4. mineral oil__________.

**B. Preparation of soap.**

5. Appearance of your soap__________.

**C. Properties of a soap.**

6. Which mixture, oil-water or oil-water-soap, forms a more stable emulsion?

7. Test tube no. 1 + $CaCl_2$__________.

8. Test tube no. 2 + $MgCl_2$__________.

9. Test tube no. 3 + tap water__________.

10. pH of your soap solution in test tube no. 4 __________.

## Post-Lab Questions

1. When you did the bromine test, which sample contained (a) the most unsaturation (–C=C–bonds) and which sample contained (b) the least unsaturation (–C=C– bonds)?

2. Does butter contain any unsaturated fatty acids? Explain your reasoning.

3. What are the two main disadvantages of soaps versus detergents?

4. Soaps that have a pH above 8.0 tend to irritate some sensitive skins. Was your soap good enough to compete with commercial preparations?

5. Explain how soap works.

# Separation of Amino Acids by Paper Chromatography

## OBJECTIVES

1. To separate amino acids and a dipeptide by paper chromatography.
2. To identify amino acids using paper chromatography.
3. To calculate $R_f$ values of amino acids.
4. To identify the hydrolysis products of aspartame using these techniques.
5. To determine the state of aspartame in a diet soft drink.

## BACKGROUND

In addition to carbohydrates and fats, proteins are an important food source. Proteins are important because they are a source for amino acids. Since not all of the amino acids required by our body can be synthesized by our body, those amino acids that cannot be synthesized, the *essential amino acids,* must be included in our diet. Why does the body require amino acids? Because amino acids are used by our body to build enzymes, tissue, and body parts, such as hair and skin. These are made up of proteins, and the amino acids are the building blocks of proteins.

All amino acids possess two functional groups in common: the carboxyl group, which gives the acidic character, and the amino group, which provides the basic character. The common structure of all amino acids is

$$\begin{array}{c} \mathrm{H} \\ | \\ R\mathrm{-C-COOH} \\ | \\ \mathrm{NH_2} \end{array}$$

The *R*-group represents the side chain that is different for each of the amino acids that are commonly found in proteins. However, all 20 amino acids have a free carboxyl group and a free amino (primary amine) group, except proline which has a cyclic side chain and a secondary amino group.

Proline

We use the properties provided by these R-groups to characterize the amino acids. The common carboxyl and amino groups provide the acid-base nature of the amino acids. The different side chains, and the solubilities provided by these side chains, can be utilized to identify the different amino acids by their rate of migration in paper chromatography. These variations in the R-group determine whether the amino acid is hydrophobic (water-fearing) or hydrophilic (water-loving), acidic or basic, or neutral.

In this experiment, we use paper chromatography to identify aspartame, an artificial sweetener, and its hydrolysis products from certain foods.

$$HOOC-CH_2-CH(NH_2)-C(=O)-NH-CH(CH_2C_6H_5)-C(=O)-OCH_3$$

Aspartame

Aspartame is the methyl ester of the dipeptide aspartylphenylalanine. Upon hydrolysis with hydrochloric acid, HCl, it yields aspartic acid, phenylalanine and methyl alcohol. When this artificial sweetener was approved by the Food and Drug Administration, opponents of aspartame claimed that it is a health hazard, because aspartame would be hydrolyzed and would yield poisonous methyl alcohol in soft drinks that are stored over long periods of time. The Food and Drug Administration ruled, however, that aspartame is sufficiently stable and fit for human consumption. Only a warning must be put on the labels containing aspartame. This warning is for patients who suffer from phenylketonurea and cannot tolerate phenylalanine.

To run a paper chromatography, we use Whatman no. 1 chromatography paper. We apply the sample (aspartame or amino acids) as a spot to a strip of chromatography paper. The paper is dipped into a mixture of solvents. The paper acts like a wick and the solvent moves up the paper by capillary action and carries the sample with it. Each amino acid may have a different migration rate depending on the solubility of the side chain in the solvent. Amino acids with similar side chains are expected to move with similar, though not identical, rates; those that have quite different side chains are expected to migrate with different velocities. Depending on the solvent system used, almost all amino acids and dipeptides can be separated from each other by paper chromatography.

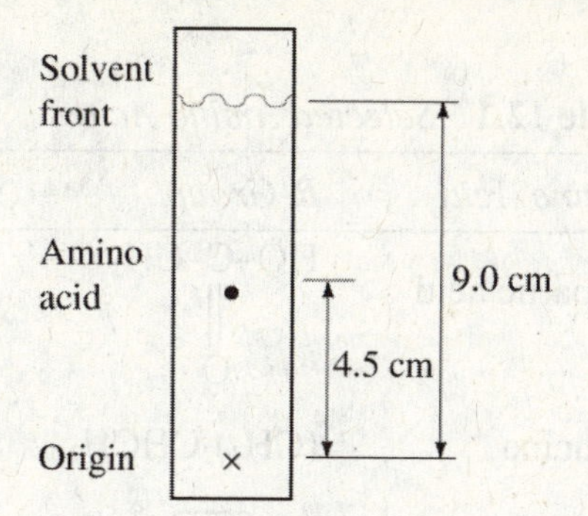


**Figure 12.1**
*Paper chromatogram.*

We actually do not measure the rate of migration of an amino acid or a dipeptide, but rather, how far a particular amino acid travels on the paper relative to the migration of the solvent. This ratio is called the $R_f$ value. In order to calculate the $R_f$ values, one must be able to visualize the position of the amino acid or dipeptide. This is done by spraying the chromatography paper with a ninhydrin solution that reacts with the amino group of the amino acid. A purple color is produced when the paper is heated. (The proline, which does not have a primary amine, gives a yellow color with ninhydrin.) For example, if the purple spot of an amino acid appears on the paper 4.5 cm away from the origin and the solvent front migrates 9.0 cm (Fig. 12.1), the $R_f$ value for the amino acid is

$$R_f = \frac{\text{Distance traveled by the amino acid}}{\text{Distance traveled by the solvent front}} = \frac{4.5\text{ cm}}{9.0\text{ cm}} = 0.50$$

In the present experiment you will determine the $R_f$ values of three amino acids: aspartic acid, leucine, and phenylalanine (Table 12.1). You will also measure the $R_f$ value of aspartame.

The aspartame you will analyze is actually a commercial sweetener, Equal®, that contains silicon dioxide, glucose, cellulose, and calcium phosphate in addition to the aspartame. None of these other ingredients of Equal® will give a purple, or any other colored spot, with ninhydrin. Other generic aspartame sweeteners may contain other non-sweetening ingredients. Occasionally, some sweeteners may contain a small amount of leucine, which can be detected by the ninhydrin test. You will also hydrolyze aspartame using HCl as a catalyst to see if the hydrolysis products will prove that the sweetener was truly aspartame. Finally, you will analyze a commercial soft drink that will be supplied by your instructor. The analysis of the soft drink can tell you if the aspartame was hydrolyzed at all during the processing and storing of the soft drink.

**Table 12.1** *Selected Amino Acids*

| *Amino Acid* | *R-Group* | *Polarity* | *Water Property* |
|---|---|---|---|
| Aspartic acid | HO–C(=O)–$CH_2$ | Polar, acidic | Hydrophilic |
| Leucine | $(CH_3)_2CHCH_2$– | Non polar, neutral | Hydrophobic |
| Phenylalanine | $C_6H_5$–$CH_2$– | Non polar, neutral | Hydrophobic |

## PROCEDURE

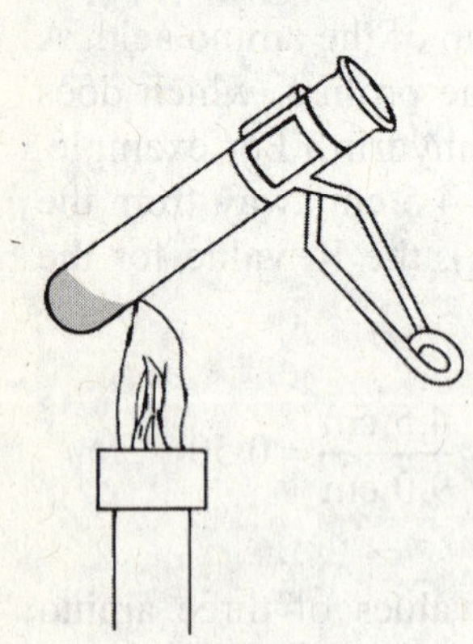

**Figure 12.2**
*Position for holding a test tube in a Bunsen flame.*

1. Dissolve about 10 mg of the sweetener Equal® in 1 mL of 6 M HCl in a test tube (13 × 100 mm). Heat it with a Bunsen burner, using a small flame (or with a microburner) to a boil for 30 sec., but make sure that the liquid does not completely evaporate. Do not heat the bottom of the test tube, but heat slightly above the surface level of the solution (Fig. 12.2). Set the solution aside to cool; label it "Hydrolyzed Aspartame."

**CAUTION!**

**Hold the test tube by a test tube holder. Heat at the top of the liquid level to minimize bumping. Point the test tube away from you and away from neighbors.**

2. Label five small test tubes (13 × 100 mm), respectively, for aspartic acid, phenylalanine, leucine, aspartame, and Diet Coca-Cola®. Place about 0.5 mL samples in each test tube.

**CAUTION!**

**Use plastic gloves throughout in order not to contaminate the paper chromatogram. Make sure you do not touch the paper with your fingers; you may hold the edges.**

3. Take a strip of Whatman no. 1 chromatographic paper, 8 × 15 cm and 0.016 thickness. With a pencil (not a pen; the ink in a pen is made up of organic dyes and the dyes will migrate like the amino acids), lightly draw a line parallel to the 8 cm edge and about 1 cm from the edge. Mark the positions of 6 spots, placed equally, where you will spot your samples (Fig.12.3).

4. Spotting. For each sample, use a separate capillary tube. Apply a drop of sample to the paper until it spreads to a spot of 1 mm diameter. Dry the spots. (If a heat lamp is available, use it for drying.) Do the spotting according to the directions shown in Figure 12.3:

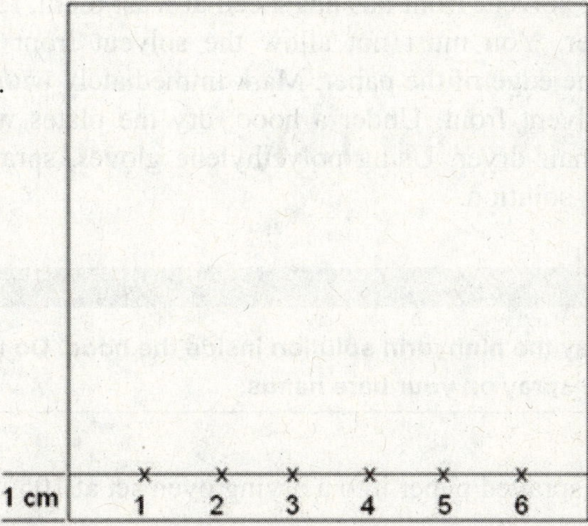


**Figure 12.3**
*Spotting. Each vertical column is called a lane.*

lane 1 - phenylalanine (1 drop)

lane 2 - aspartic acid (1 drop)

lane 3 - leucine (1 drop)

lane 4 - aspartame (in Equal®) (1 drop)

lane 5 - hydrolyzed aspartame (6 drops)

lane 6 - Diet Coca Cola® (15 drops)

When you do multiple drops, as in lanes 5 and 6, allow the paper to dry after each drop before applying the next drop. Do not allow the spots to spread to larger than 2 mm in diameter.

5. Pour about 10 mL of solvent mixture (butanol:acetic acid:water) into a 1 L beaker. Keep the beaker covered with aluminum foil as soon as you pour in the solvent and throughout the experiment. While still wearing gloves, curl the dry paper into a cylinder with the spots on the outside and staple it at either end of the paper to hold it together (Fig. 12.4). Place the paper in the chromatography tank (the 1 L beaker) so that **the spots are ABOVE the level of the solvent** (Fig.12.5).

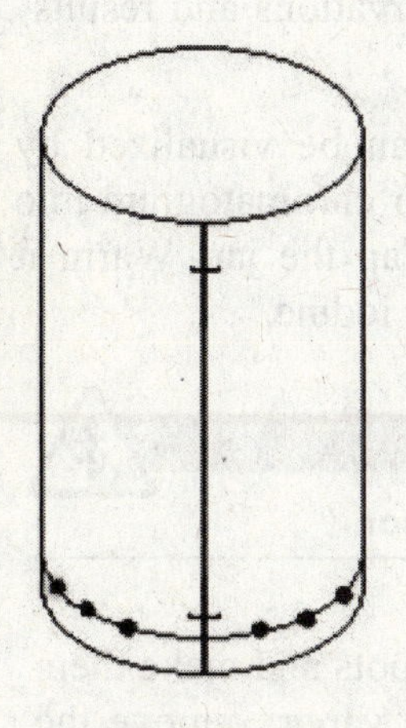

**Figure 12.4**
*Curl and staple the chromatography paper.*

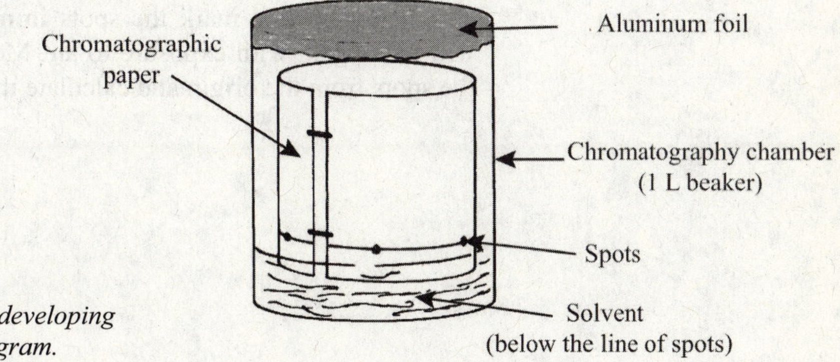


**Figure 12.5**
*Chamber for developing the chromatogram.*

6. Take the covered beaker with the paper inside and carefully place it on a hot plate. With a low setting (for example, no. 3 out of 10), heat the beaker to about 35 °C. Allow the solvent front to advance at least 6 cm (it may take 40–50 min.), but do not allow it to get closer than 1 cm from the upper edge.

7. When the solvent front has advanced at least 6 cm, remove the paper from the beaker. You must not allow the solvent front to advance up to or beyond the edge of the paper. Mark immediately *with a pencil* the position of the solvent front. Under a hood, dry the plates with the aid of a heat lamp or hair dryer. Using polyethylene gloves, spray the dry paper with ninhydrin solution.

**CAUTION!**

**You must spray the ninhydrin solution inside the hood. Do not breathe the fumes. Do not get the spray on your bare hands.**

8. Place the sprayed paper into a drying oven set at 105–110 °C for 2–3 min.

9. Remove the sprayed paper from the oven. Mark the center of the spots and calculate the $R_f$ values of each spot. Record your observations and results on the Report Sheet.

10. If the spots on the chromatogram are faded, they can be visualized by exposing the chromatogram to iodine vapor. Place your chromatogram into a wide-mouth jar containing a few iodine crystals. Cap the jar. Warm it **slightly** on a hot plate to enhance the sublimation of the iodine.

**CAUTION!**

**Heat in the hood. Iodine vapor is toxic. Do not breathe the vapor.**

The iodine vapor will interact with the faded pigment spots and make them visible. After a few minutes of exposure to the iodine vapor, remove the chromatogram and mark the spots **immediately** with a pencil. The spots will fade again with exposure to air. Measure the distance of the center of the spots from the origin and calculate the $R_f$ values.

## CHEMICALS AND EQUIPMENT

1. 0.12% solutions of aspartic acid, phenylalanine, and leucine
2. 0.5% solution of aspartame (Equal®)
3. Diet Coca-Cola®
4. 6 M HCl
5. 0.2% ninhydrine spray
6. Butanol:acetic acid:water, solvent mixture
7. Equal® sweetener
8. Aluminum foil
9. Whatman no. 1 chromatographic paper, 15 × 8 cm
10. Ruler
11. Polyethylene gloves
12. Capillary tubes, open on both ends
13. Heat lamp or hair dryer
14. Drying oven ,110 °C
15. Wide-mouth jar
16. Iodine crystals
17. Test tubes (13 × 100 mm)
18. 1-L beaker

name (print) | date (of lab meeting) | grade

course/section | partner's name (if applicable)

12

## EXPERIMENT 12: SEPARATION OF AMINO ACIDS BY PAPER CHROMATOGRPAHY

# Prelab Questions

**A. Safety concerns**

1. What would happen if you didn't use gloves and your finger came into contact with the ninhydrin spray?

2. How do you handle a test tube when heating with a Bunsen burner?

3. Why use iodine in the hood?

**B. Basic principles.**

1. If an amino acid moved 5.0 cm on the paper and the solvent moved 7.0 cm, what is the $R_f$ value of the amino acid?

2. Why must you use a pencil and not ink to mark the origin of your spot on the paper?

3. What functional groups are present in an amino acid?

name (print) | date (of lab meeting) | grade

course/section | partner's name (if applicable)

12

## EXPERIMENT 12: SEPARATION OF AMINO ACIDS BY PAPER CHROMATOGRPAHY

# *Report Sheet*

1.

| *Sample* | *Distance traveled (mm)* | *Solvent front (mm)* | $R_f$ |
|---|---|---|---|
| Phenylalanine | | | |
| Aspartic acid | | | |
| Leucine | | | |
| Aspartame (in Equal®) | | | |
| Hydrolyzed aspartame | | | |
| Diet Coca-Cola® | | | |

2. Identification

   (a) Name the amino acids you found in the hydrolysate of the sweetener, Equal®.

   (b) How many spots were stained with ninhydrin (1) in Equal® and (2) in Diet Coca-Cola® samples?

## Post-Lab Questions

1. What effect does the R-group have on the properties of an amino acid?

2. In testing the hydrolysate of aspartame, you forgot to mark the position of the solvent front on your TLC plate.

   (a) Could you still determine how many amino acids were in the aspartame?

   (b) Could you still identify those amino acids?

3. Do you have any evidence that the aspartame was hydrolyzed during the processing and storage of the Diet Coca-Cola® sample? Explain.

4. Aspartic acid moves slower on the chromatographic paper than either leucine or phenylalanine. How do you account for the difference?

EXPERIMENT 13

# Isolation and Identification of Casein

## OBJECTIVES

1. To precipitate protein at isoelectric conditions.
2. To isolate casein from milk.
3. To perform some chemical tests to characterize proteins.

## BACKGROUND

Casein is the most important protein in milk. It functions as a storage protein, fulfilling nutritional requirements. Casein can be isolated from milk by acidification which brings the protein to its isoelectric point. At the isoelectric point the number of positive charges on a protein equal the number of negative charges. Proteins are least soluble in water at their isoelectric points because they tend to aggregate by electrostatic interaction. The positive end of one protein molecule attracts the negative end of another protein molecule, and the aggregates precipitate out of solution.

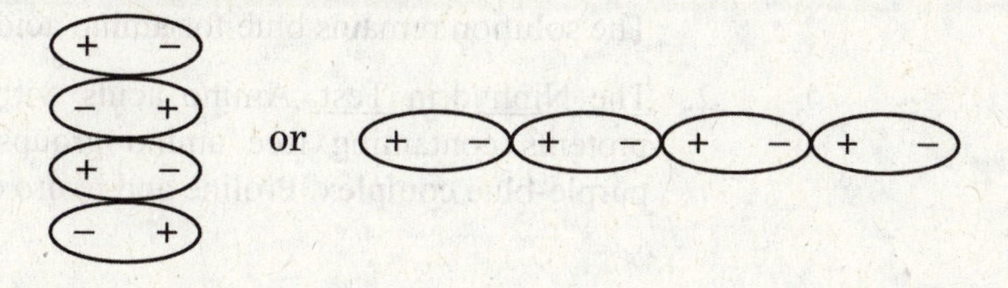


On the other hand, if a protein molecule has a net positive (at low pH or acidic condition) or a net negative charge (at high pH or basic condition), its solubility in water is increased.

$$\overset{+}{N}H_3 \sim COOH \xleftarrow[\text{low pH}]{H^+} \overset{+}{N}H_3 \sim COO^- \xrightarrow[\text{high pH}]{OH^-} NH_2 \sim COO^- + H_2O$$

More soluble | Least soluble (at isoelectric pH) | More soluble

In the first part of this experiment you are going to isolate casein from milk which has a pH of about 7. Casein will be separated as an insoluble precipitate by acidification of the milk to its isoelectric point (pH = 4.6). The fat that precipitates along with casein can be removed by dissolving it in alcohol.

In the second part of this experiment you are going to show that the precipitated milk product is a protein. The identification will be achieved by performing a few important chemical tests.

1. <u>The Biuret Test.</u> This is one of the most general tests for proteins. When a protein reacts with copper(II) sulfate, a positive test is the formation of a copper complex which has a violet color.

$$(-\overset{\overset{\displaystyle O}{\|}}{C}-NH-)_n + Cu^{2+} \longrightarrow \text{[Cu complex: Cu coordinated by four }N\text{ and two }H_2O\text{]}$$

Protein — Blue color — Protein–copper complex (violet color)

This test works for any protein or compound that contains two or more of the following groups:

$$-\overset{\overset{\displaystyle O}{\|}}{C}-NH-,\ -\overset{\overset{\displaystyle O}{\|}}{C}-NH_2,\ -CH_2-NH_2,\ -\overset{\overset{\displaystyle NH}{\|}}{C}-NH_2,\ -\overset{\overset{\displaystyle S}{\|}}{C}-NH_2$$

The solution remains blue for amino acids and dipeptides.

2. <u>The Ninhydrin Test.</u> Amino acids with a free amino group, $-NH_2$, and proteins containing free amino groups react with ninhydrin to give a purple-blue complex. Proline and hydroxyproline form a yellow color.

$$\underset{\text{Amino acid}}{NH_2-\underset{\underset{\displaystyle R}{|}}{CH}-COOH} + 2\ \underset{\text{Ninhydrin}}{\text{[ninhydrin]}} \longrightarrow \underset{\text{Purple-blue complex}}{\text{[indanedione}-N=\text{indanedione]}} + RCHO + CO_2 + 3H_2O$$

3. Heavy Metal Ions Test. Heavy metal ions precipitate proteins from solution. The ions that are most commonly used for protein precipitation are $Zn^{2+}$, $Fe^{3+}$, $Cu^{2+}$, $Sb^{3+}$, $Ag^{+}$, $Cd^{2+}$, and $Pb^{2+}$. Among these metal ions, $Hg^{2+}$, $Cd^{2+}$, and $Pb^{2+}$ are known for their notorious toxicity to humans. They can cause serious damage to proteins (especially enzymes) by denaturing them. This can result in death. The precipitation occurs because proteins become cross-linked by heavy metals as shown below:

$$2NH_2\sim\sim\sim C(=O)-O^- + Hg^{2+} \longrightarrow \left\{ \begin{matrix} H_2N \rightarrow & Hg & \leftarrow NH_2 \\ C(=O)-O & & O-C(=O) \end{matrix} \right\}$$

Insoluble precipitate

$$2 \;{>}SH + Hg^{2+} \longrightarrow {>}S^- \, Hg^{2+} \, {}^-S{<}$$

Insoluble precipitate

Victims swallowing $Hg^{2+}$ or $Pb^{2+}$ ions are often treated with an antidote of a food rich in proteins, which can combine with mercury or lead ions in the victim's stomach and, hopefully, prevent absorption. Milk and raw egg white are used most often. The insoluble complexes are then immediately removed from the stomach by an emetic, an agent that induces vomiting.

4. The Xanthoprotein Test. This is a characteristic reaction of proteins that contain phenyl rings.

$$C_6H_5-$$

Concentrated nitric acid reacts with the phenyl ring to give a yellow-colored aromatic nitro compound. Addition of alkali at this point will deepen the color to orange.

$$HO-C_6H_4-CH_2-CH(NH_2)-COOH + HNO_3 \longrightarrow HO-C_6H_3(NO_2)-CH_2-CH(NH_2)-COOH + H_2O$$

Tyrosine      Colored compound

The yellow stains on the skin caused by nitric acid are the result of the xanthoprotein reaction.

## PROCEDURE

**Isolation of Casein**

1. To a 250-mL Erlenmeyer flask, add 50 g of milk (determine and record the mass to the nearest 0.01 g) and heat the flask in a water bath (a 600-mL beaker containing about 200 mL of tap water; see Fig. 13.1). Stir the solution constantly with a stirring rod. When the bath temperature has reached about 40 °C, remove the flask from the water bath and add about ten drops of glacial acetic acid while stirring. Observe the formation of a precipitate.

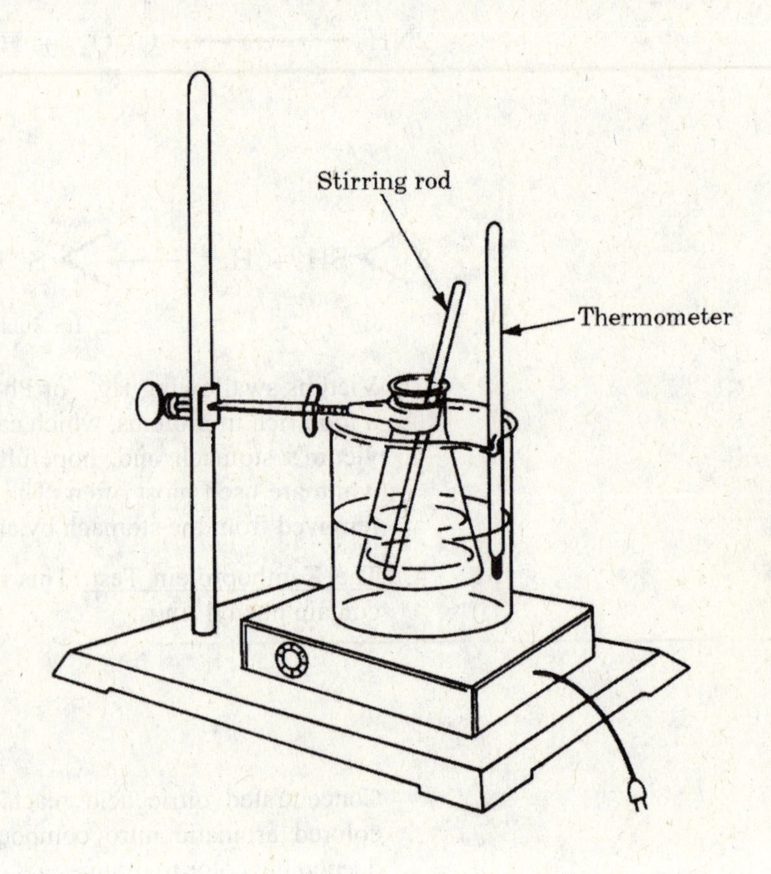


**Figure 13.1**
*Precipitation of casein.*

2. Filter the mixture into a 100-mL beaker by pouring it through a cheese cloth which is fastened with a rubber band over the mouth of the beaker (Fig. 13.2). Remove most of the water from the precipitate by squeezing the cloth gently. Discard the filtrate in the beaker. Using a spatula, scrape the precipitate from the cheese cloth into the empty flask.

3. Add 25 mL of 95% ethanol to the flask. After stirring the mixture for 5 min., allow the solid to settle. Carefully decant (pour off) the liquid that contains fats into a beaker. Discard the liquid.

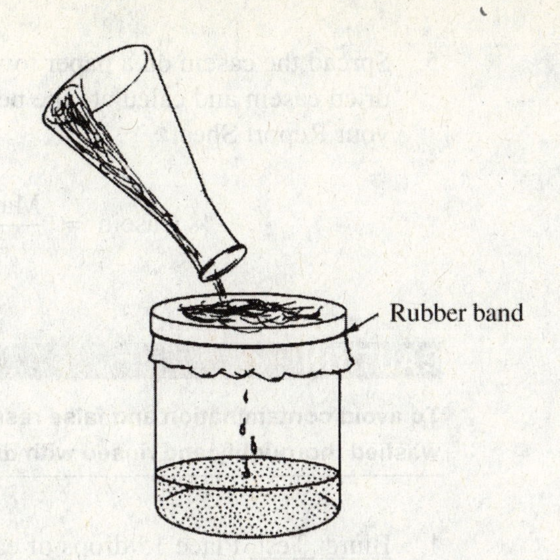


**Figure 13.2**
*Filtration of casein.*

4. To the residue, add 25 mL of a 1:1 mixture of diethyl ether:ethanol. After stirring the resulting mixture for 5 min, collect the solid by vacuum filtration with a Büchner funnel (Fig. 13.3).

**CAUTION!** 

**Diethyl ether is highly flammable. Make sure there is no open flame in the laboratory.**

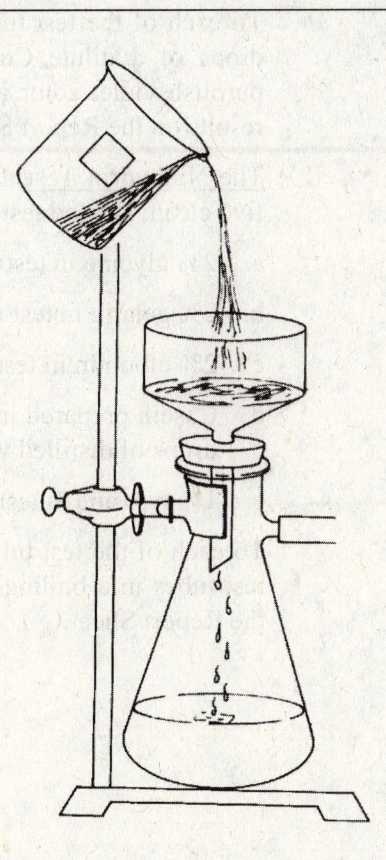

**Figure 13.3**
*Vacuum filtration with a Büchner funnel.*

5. Spread the casein on a paper towel and let it dry. Determine the mass of the dried casein and calculate the percentage of casein in the milk. Record it on your Report Sheet.

$$\% \text{ Casein} = \frac{\text{Mass of solid (casein), g}}{\text{Mass of milk, g}} \times 100$$

**CAUTION!**

**To avoid contamination and false results, it is important that all test tubes are washed thoroughly and rinsed with distilled water.**

**Chemical Analysis of Proteins**

1. Biuret Test. Place 15 drops of each of the following solutions in five clean, labeled test tubes (13 × 100 mm);
   a. 2% glycine in test tube no. 1
   b. 2% gelatin in test tube no. 2
   c. 2% albumin in test tube no. 3
   d. Casein prepared in Part A (one quarter of a full micro-spatula) plus 15 drops of distilled water in test tube no. 4
   e. 1% tyrosine in test tube no. 5

   To each of the test tubes, add five drops of 10% NaOH solution and two drops of a dilute $CuSO_4$ solution while swirling. The development of purplish violet color is evidence of the presence of proteins. Record your results on the Report Sheet.

2. The Ninhydrin Test. Place 15 drops of each of the following solutions in five clean, labeled test tubes (13 × 100 mm):
   a. 2% glycine in test tube no. 1
   b. 2% gelatin in test tube no. 2
   c. 2% albumin in test tube no. 3
   d. Casein prepared in Part A (one quarter of a full micro-spatula) plus 15 drops of distilled water in test tube no. 4
   e. 1% tyrosine in test tube no. 5

   To each of the test tubes, add five drops of ninhydrin reagent and heat the test tubes in a boiling water bath for about 5 min. Record your results on the Report Sheet.

3. <u>Heavy Metal Ions Test.</u> Place 2 mL of milk in each of three clean, labeled test tubes (13 × 100 mm). Add a few drops of each of the following metal ions to the corresponding test tubes as indicated below:

   a. $Pb^{2+}$ as $Pb(NO_3)_2$ in test tube no. 1

   b. $Hg^{2+}$ as $Hg(NO_3)_2$ in test tube no. 2

   c. $Na^+$ as $NaNO_3$ in test tube no. 3

   Record your results on the Report Sheet.

**CAUTION!**

**Since the following test involves the use of concentrated nitric acid, the acid will be dispensed by your instructor in the hood. Wear gloves and goggles.**

4. <u>The Xanthoprotein Test.</u> Place 15 drops of each of the following solutions in five clean, labeled test tubes (13 × 100 mm):

   a. 2% glycine in test tube no. 1

   b. 2% gelatin in test tube no. 2

   c. 2% albumin in test tube no. 3

   d. Casein prepared in Part A (one quarter of a full micro-spatula) plus 15 drops of distilled water in test tube no. 4

   e. 1% tyrosine in test tube no. 5

   Ten drops of concentrated $HNO_3$ are added to each test tube and swirled. Heat the test tubes carefully in a warm water bath at 60 °C. Observe any change in color. Record the results on your Report Sheet.

## CHEMICALS AND EQUIPMENT

1. Hot plate
2. Büchner funnel in no. 7 one-hole rubber stopper
3. 500-mL filter flask
4. Filter paper (Whatman no. 2, 7 cm)
5. Cheese cloth
6. Rubber band
7. Boiling chips
8. 95% ethanol
9. Diethyl ether-ethanol mixture
10. Regular milk
11. Glacial acetic acid
12. Concentrated nitric acid
13. 2% albumin
14. 2% gelatin
15. 2% glycine
16. 5% copper(II) sulfate
17. 5% lead(II) nitrate
18. 5% mercury(II) nitrate
19. Ninhydrin reagent
20. 10% sodium hydroxide
21. 1% tyrosine
22. 5% sodium nitrate
23. Test tubes (13 × 100 mm)
24. 250-mL Erlenmeyer flask
25. 100-mL beaker

name (print) date (of lab meeting) grade

course/section partner's name (if applicable)

13 **EXPERIMENT 13: ISOLATION AND IDENTIFICATION OF CASEIN**

# *Prelab Questions*

**A. Safety concerns.**

1. Why are flames in the laboratory a concern in this experiment?

2. Why is it important to wash and rinse all test tubes before carrying out any tests?

3. What precautions do you take when dealing with concentrated nitric acid, and why should you be careful?

**B. Basic principles.**

1. What function does the casein play in milk?

2. Why are proteins least soluble at their isoelectric point?

3. What structural group must be present in an amino acid in order to give a positive xanthoprotein test?

4. What are the three most toxic heavy metal ions?

name (print) ______ date (of lab meeting) ______ grade ______

course/section ______ partner's name (if applicable) ______

13 **EXPERIMENT 13: ISOLATION AND IDENTIFICATION OF CASEIN**

# *Report Sheet*

**Part A. Isolation of casein**

1. Mass of milk ______ g

2. Mass of dried casein ______ g

3. Percentage of casein in milk ______ %

Show your calculations.

**Part B. Chemical analysis of proteins**

***Biuret Test***

| *Test Tube No.* | *Substance* | *Color Formed* |
|---|---|---|
| 1 | 2% glycine | |
| 2 | 2% gelatin | |
| 3 | 2% albumin | |
| 4 | Casein + $H_2O$ | |
| 5 | 1% tyrosine | |

Which of these chemicals gives a positive test with this reagent? ______

***Ninhydrin Test***

| *Test Tube No.* | *Substance* | *Color Formed after Heating* |
|---|---|---|
| 1 | 2% glycine | |
| 2 | 2% gelatin | |
| 3 | 2% albumin | |
| 4 | Casein + $H_2O$ | |
| 5 | 1% tyrosine | |

Which of these chemicals gives a positive test with this reagent? ____________

***Heavy Metal Ion Test***

| *Test Tube No.* | *Substance* | *Precipitates Formed* |
|---|---|---|
| 1 | $Pb(NO_3)_2$ | |
| 2 | $Hg(NO_3)_2$ | |
| 3 | $NaNO_3$ | |

Which of these metal ions gives a positive test with casein in milk? ____________

***Xanthoprotein Test***

| *Test Tube No.* | *Substance* | *Color before Heating* | *Color after Heating* |
|---|---|---|---|
| 1 | 2% glycine | | |
| 2 | 2% gelatin | | |
| 3 | 2% albumin | | |
| 4 | Casein + $H_2O$ | | |
| 5 | 1% tyrosine | | |

Which of these chemicals gives a positive test with this reagent? ____________

## Post-Lab Questions

1. Explain why casein precipitates when acetic acid is added to it.

2. How is the milk fat removed in the isolation procedure?

3. What functional group(s) will give a positive reaction with the ninhydrin reagent?

4. If by mistake (**DO NOT TRY IT!**) your finger touched concentrated nitric acid and you observed a yellow color on your fingers, what functional group(s) in your skin is (are) responsible for this reaction?

5. Why is milk or raw egg used as an antidote in cases of heavy metal ion poisoning?

6. A glass of milk has a mass of about 175 g. Using your results, how many grams of casein would you expect to find in that glass of milk?

# Enzymes

## OBJECTIVES

1. To demonstrate the role of enzymes as catalysts.
2. To observe how concentration, temperature, pH, and inhibitors effect enzyme activity.

## BACKGROUND

Each cell in our body operates like a chemical factory. Chemicals are broken down for raw material and energy, and new chemicals are synthesized. Our diet supplies only a few of the many compounds required for the operation of our body. Most are synthesized within the cell by the hundreds of different types of reactions as part of metabolism. What is remarkable is that these reactions would not be able to take place at body temperature, or if they were able to take place at all, they would occur at an extremely slow rate. What allows these reactions to be rapid and efficient are the chemical compounds called *enzymes*.

Enzymes are large protein molecules that behave as *catalysts*. A catalyst is a chemical agent that allows a chemical reaction to go faster without itself undergoing any change - it takes part in the reaction but looks the same after the reaction is done. Since it looks the same at the end of a reaction, it can be used over and over, again and again, in chemical reactions. Think about the difficulty we would experience if food would take weeks to be digested and our muscles and nerves would not work as they do. Our life as we know it would not be the same.

Like any catalyst, enzymes do not alter the position of an equilibrium. If a reaction would not take place, an enzyme would not make it happen. Enzymes influence rates only; in their presence, rates take place faster. They do this by lowering the energy of activation (Fig. 14.1).

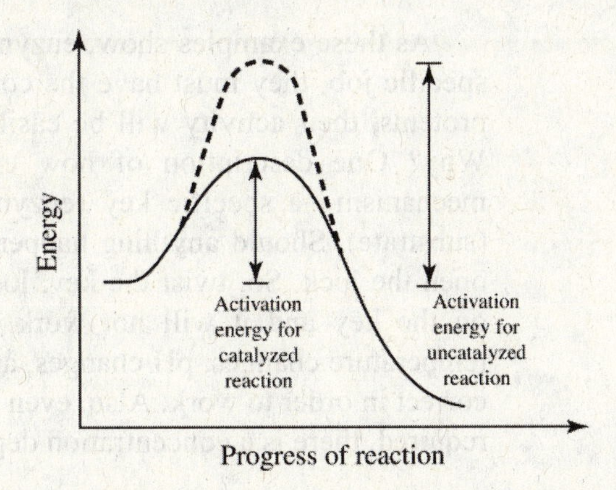


**Figure 14.1**
*Effect of an enzyme on the energy of activation.*

The energy of activation is the energy necessary for bond-breaking and bond-forming processes to take place. It is the barrier over which molecules must make for a reaction to occur. The effectiveness of enzymes is remarkable–depending on the reaction and the enzyme, rates can be from $10^9$ to $10^{20}$ times faster than the uncatalyzed reaction under the same set of conditions.

Another aspect of enzymes makes them remarkable. Within each cell there are 2000 to 3000 different enzymes. Depending on the type of cell, there are different sets of enzymes. However, since enzymes are composed of proteins, these large molecules are highly specific–*an enzyme will catalyze one specific reaction with one specific substrate*. For example, the enzyme urease catalyzes only the hydrolysis of urea and not that of any other amides, even though they may be closely related.

$$(NH_2)_2C{=}O + H_2O \xrightarrow{\text{urease}} 2NH_3 + CO_2$$

Catylase or peroxidase is an enzyme found in most cells and serves to break down peroxides produced from some metabolic processes. You probably have seen its effects if you ever applied hydrogen peroxide to an open cut. The white foam produced is due to the action of the enzyme on the hydrogen peroxide releasing oxygen; the reaction is faster than without the catalyst.

$$2H_2O_2 \xrightarrow{\text{peroxidase}} 2H_2O + O_2$$

For another example, let's consider amylase in some detail. This enzyme is found in human saliva and has a shape that can recognize the polysaccharide amylose (starch). It helps to hydrolyze α-1,4-glycosidic linkages in amylose into the smaller sugar units, the disaccharide molecule, maltose. It tends to work best in the mouth where the pH is neutral to slightly alkaline. But when you swallow and the food gets to the stomach, it stops working since the stomach pH of 2 is too acidic for it. However, the food eventually gets to the gut where the pancreas secretes more amylase and the hydrolysis continues to produce maltose. The enzyme maltase then converts the maltose into glucose which crosses the intestinal barrier and enters the blood.

$$\text{Amylose} \xrightarrow{\text{amylase}} \text{Oligosaccharides} \xrightarrow{\text{amylase}} \text{Maltose} \xrightarrow{\text{maltase}} \text{Glucose}$$

As these examples show, enzymes are highly specific. In order to do their specific job, they must have the correct shape. Since enzymes are made from proteins, their activity will be easily affected by heat, pH, and heavy metals. Why? One description of how enzymes work is by the "lock and key" mechanism—a specific key (enzyme) is required in order to open the lock (substrate). Should anything happen to the shape of the key, it no longer will open the lock. So, twist the key, lose a tooth, or have some foreign substance on the key and it will not work in the lock. Enzymes become altered by temperature changes, pH changes, and metal inhibitors. The conditions must be correct in order to work. Also, even though small amounts of these catalysts are required, there is a concentration dependency.

## Factors Affecting Enzyme Activity

### A. Concentration

In order for an enzyme to show its catalytic effect, the enzyme must combine with the starting substrate. The resulting enzyme-substrate complex comes about when there is a proper fit at the place of the reaction, the *active site*. Products are released and the enzyme is then recycled for another reaction. Schematically, it can be represented by the following:

$$\text{Enzyme} + \text{Substrate} \rightleftharpoons \text{Enzyme–substrate complex} \rightleftharpoons \text{Enzyme} + \text{Products}$$

$$\text{E} + \text{S} \rightleftharpoons \text{E-S}^{\ddagger} \rightleftharpoons \text{E} + \text{P}$$

Since the enzyme is a catalyst, its concentration is usually very much smaller than that of the substrate. In this model, if the substrate concentration is held constant, any increase in the enzyme concentration will result in a rate increase. In fact, there is a direct relationship, so that the rate increases linearly (Fig. 14.2): double the enzyme concentration and the rate of the reaction increases by a factor of two.

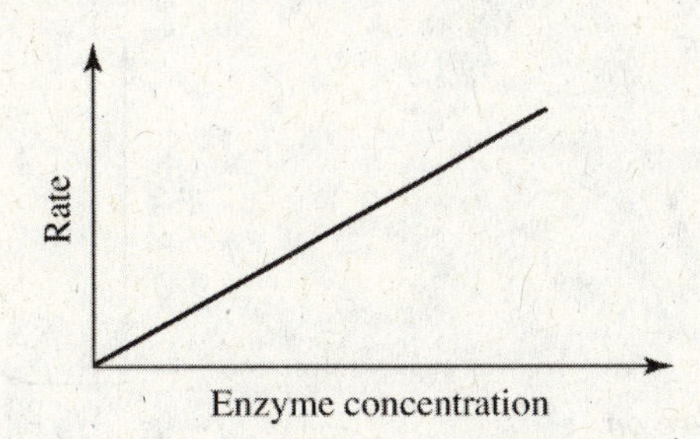


**Figure 14.2**
*The effect of enzyme concentration on the rate of an enzyme-catalyzed reaction: a linear curve.*

There is another effect of concentration. Suppose the concentration of the enzyme is held constant this time. If substrate concentration is initially low and is then increased, the rate will also increase. However, this initial increase will not be continuous. As enzyme and substrate combine and release products, more and more active sites become occupied. Eventually, as substrate concentration increases, all available active sites become occupied, and further increases in substrate concentration will no longer show a rate increase. The reaction is proceeding at its maximum rate and the rate stays the same; enzymes are cycling as fast as possible and excess substrate cannot find any active sites to which to attach. We have reached saturation (Fig. 14.3).

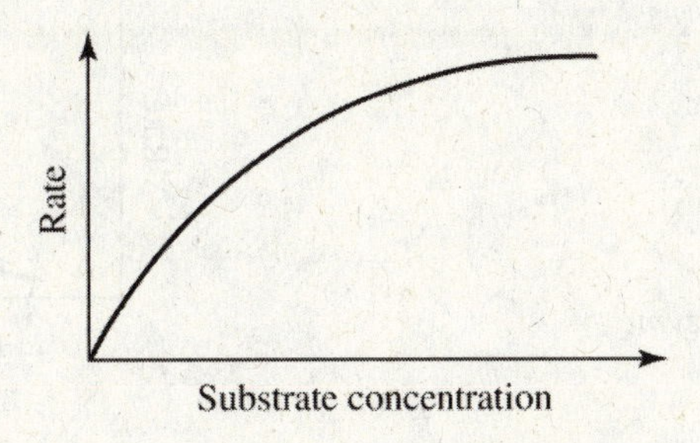


**Figure 14.3**
*The effect of substrate on the rate of an enzyme-catalyzed reaction: a saturation curve.*

### B. Temperature

In an earlier experiment (see Expt. 5), we saw that the rate of a reaction increased as the temperature increased. This was for an uncatalyzed reaction. With a reaction that is catalyzed by an enzyme, the effect of temperature takes a different course. If the temperature is low, any increase in temperature will cause an increase in the rate. This is what is normally expected by the kinetic-molecular theory - more effective collisions between enzyme molecules and substrate molecules leads to more product. However, once an optimal temperature is reached, the rate is at a maximum and the enzyme shows its greatest efficiency. Any further increase in temperature will lead to a decrease in the rate (Fig. 14.4). The decline comes about because temperature affects enzyme conformation. Enzyme and substrate no longer fit together properly, so the rate leading to product falls. Consider some practical application of these effects: (1) deactivation of enzymes at low temperature is the reason refrigeration works in the preservation of food; (2) sterilization in an autoclave works since bacterial proteins and enzymes are denatured at elevated temperatures and are deactivated.

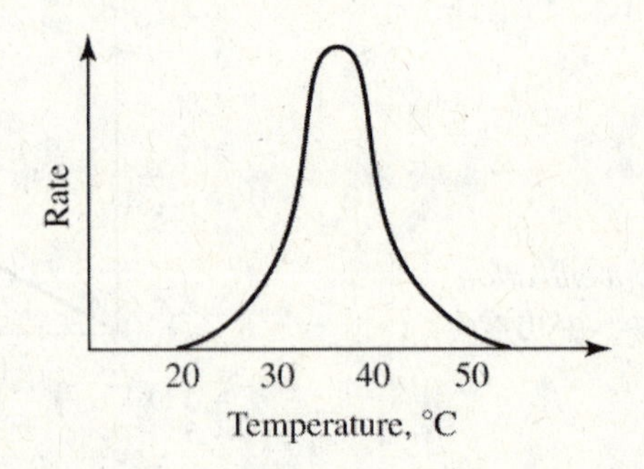


**Figure 14.4**
*The effect of temperature on enzyme activity.*

### C. pH

Changes in pH also have an effect on enzyme activity. As with temperature, pH can alter the conformation of proteins and thus, enzymes. Each enzyme operates at an optimal pH (Fig. 14.5) and is most active at that pH. For example, pepsin, an enzyme found in the stomach, functions at a pH of 2; trypsin, found in the gut, operates efficiently at a pH of 8. Deviations from these optimum values leads to deactivation, often due to denaturation.

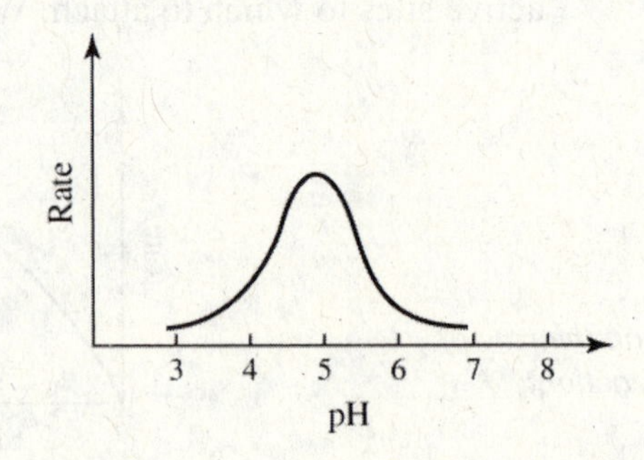


**Figure 14.5**
*The effect of pH on enzyme activity.*

### D. Inhibitors

Any substance which reduces or stops an enzyme from behaving as a catalyst is an *inhibitor*. If the substance binds to the active site (the site where the substrate usually combines with the enzyme during reaction), then the substrate is prevented from entering the active site. The inhibitor competes with the substrate for the enzyme. This *competitive inhibitor* causes the enzyme to lose its activity. Some inhibitors bind to some other portion of the enzyme structure and cause a change in the shape of the active site, and the substrate cannot bind. This *noncompetitive inhibitor* also causes enzyme activity to decrease. Poisons, toxins, heavy metals, and some drugs behave either as a competitive or noncompetitive inhibitor. The net result is a loss in enzyme activity.

## PROCEDURE

**General**

You will need three water baths for this experiment. Half-fill three 400-mL beakers. Keep one at 0–5 °C with ice-water; keep a second at 35–40 °C and monitor the temperature of this one carefully; keep the third one at boiling, 100 °C.

**CAUTION!**

**Wear safety glasses throughout this experiment. Handle hot beakers with beaker tongs.**

**Catalysis**

1. Take three clean, dry test tubes (13 × 100 mm) and label them no. 1, no. 2, and no. 3, respectively. Add 2 mL 3% $H_2O_2$ (hydrogen peroxide) to each test tube.
2. To test tube no. 1, add a small, pea-sized piece of raw liver. To test tube no. 2, add a small, pea-sized piece of cooked liver. Add nothing to test tube no. 3.
3. Place all three test tubes into the water bath held at 35–40 °C.
4. Which solution quickly produces foam? Record your observation on the Report Sheet (1). What accounts for the difference?

**Enzyme Concentration**

1. Take three clean, dry test tubes (13 × 100 mm) and label them no. 1, no. 2, and no. 3.
2. Prepare solutions according to the table following.

| *Test Tube No.* | *2% Rennin Solution, mL* | *Distilled Water, mL* | *Milk, mL* |
|---|---|---|---|
| 1 | 4 | 0 | 3 |
| 2 | 2 | 2 | 3 |
| 3 | 1 | 3 | 3 |

3. Add 1 mL 3 M HCl (hydrochloric acid) to each test tube. Mix by tapping with your finger.
4. At the same time, place all three test tubes into the water bath at 35–40 °C.
5. With a stop-watch or seconds-hand of a watch, keep track of the time the test tubes are in the bath. Note the order in which the solutions form a precipitate. Which test tube solution forms a precipitate quickest? How long did it take? What about the others? Record your observations on the Report Sheet (2).

**Temperature**

1. Collect 6 mL of your own saliva; saliva contains salivary amylase. Use a 10-mL graduated cylinder to collect your saliva. Until you use the saliva, keep the graduated cylinder in the water bath at 35–40 °C.

**CAUTION!** 

**Use gloves when collecting and handling the saliva, even your own. Pathogens may be present in any body fluid.**

2. Place 2 mL of 2% starch solution in each of three clean, dry test tubes (13 × 100 mm) labeled no. 1, no. 2, and no. 3. Into each test tube, add 2 mL of saliva and mix. Place test tube no. 1 into the ice-water bath; place test tube no. 2 into the water bath at 35–40 °C; place test tube no. 3 into the boiling water bath. Allow the test tubes with their contents to stand in their respective baths for at least 30 min. (While these are heating, you can work on the next section.)
3. After heating for 30 min., transfer 3 drops of each solution into separate depressions of a white spot plate. Use clean, separate medicine droppers for the transfers. To each sample, add 2 drops of iodine solution. (For reference, test starch in a separate depression, and only the iodine solution in another.)
4. Record the color of the solutions on your Report Sheet and estimate the extent of the hydrolysis by the color of the test (3). [Starch can form an intense, brilliant, dark blue or violet colored complex with iodine. The straight chain component of starch, the amylose, gives a blue color while the branched component, the amylopectin yields a purple color. In the presence of iodine the amylose forms helixes inside of which the iodine molecule assemble as long polyiodide chains. The helix forming branches of amylopectin are much shorter than those of amylose. Therefore, the polyiodide chains are also much shorter in the amylopectin-iodine complex than in the amylose-iodine complex. The result is a different color (purple). When starch is hydrolyzed and broken down to small carbohydrate units (oligosaccharides) the iodine will not give a dark blue (or purple) color; the color will vary in shades of brown. The iodine test is used in this experiment to indicate the degree of completion of the hydrolysis.]

**pH**

1. Take three clean, dry test tubes (16 × 125 mm) and label them no. 1, no. 2, and no. 3, respectively. Place 2 mL of buffer solution of pH 2 into test tube no.1; place 2 mL of buffer solution of pH 7 into test tube no. 2; place 2 mL of buffer solution of pH 12 into test tube no. 3.
2. Into each test tube add 2 mL amylase solution (or saliva) and 2 mL 1% starch solution. Mix and place the test tubes into the water bath at 35–40°C. Heat for at least 30 min. (While these are heating, you can work on the next section.)
3. After heating for 30 min., transfer 3 drops of each solution into separate depressions of a white spot plate. Use clean, separate medicine droppers for the transfers. To each sample, add 2 drops of iodine solution. (For reference, test starch in a separate depression and only the iodine solution in another.)
4. Record the color of the solutions on your Report Sheet and estimate the extent of the hydrolysis by the color of the test (4).

**Inhibitor**

1. Take a clean, dry test tube (16 × 125 mm) and add to it 2 mL amylase solution (or saliva), 2 mL 1% starch, and 10 drops of 0.1 M $Pb(NO_3)_2$.
2. Mix and place the test tube into the water bath at 35–40°C. Heat for at least 30 min.
3. After heating for 30 min., transfer 3 drops of solution into a depression of a white spot plate. Use a clean, medicine dropper for the transfer. To the sample, add 2 drops of iodine solution.
4. Record the color of the solution on your Report Sheet and estimate the extent of the hydrolysis by the color of the test (5).

## CHEMICALS AND EQUIPMENT

1. Hot plates
2. Medicine droppers
3. Stop watch
4. White spot plate
5. Amylase solution
6. Buffer, pH 2
7. Buffer, pH 7
8. Buffer, pH 12
9. 3 M HCl
10. 3% $H_2O_2$, hydrogen peroxide
11. Liver
12. Milk
13. 2% rennin solution
14. 1% starch solution
15. 2% starch solution
16. 0.01 M iodine in KI
17. 0.1 M $Pb(NO_3)_2$
18. Test tube (16 × 125 mm)
19. Test tubes (13 × 100 mm)
20. 400-mL beaker
21. 10-mL graduated cylinder

name (print) | date (of lab meeting) | grade

course/section | partner's name (if applicable)

14 **EXPERIMENT 14: ENZYMES**

# *Prelab Questions*

**A. Safety concerns**

1. Why do you wear your safety glasses for this experiment?

2. How do you handle hot glassware, such as a hot beaker?

3. Why use gloves in this experiment?

**B. Basic principles**

1. What is the role of an enzyme?

2. List at least three factors that affect the activity of an enzyme?

3. Reactions take place at what part of the enzyme?

4. Why does an enzyme lose activity at high temperatures?

5. Can an enzyme function at any pH? Why or why not?

name (print) | date (of lab meeting) | grade

course/section | partner's name (if applicable)

14 **EXPERIMENT 14: ENZYMES**

# *Report Sheet*

## 1. Effect of a Catalyst

| *Test Tube No.* | *Observation* | *Conclusion* |
|---|---|---|
| 1 | | |
| 2 | | |
| 3 | | |

## 2. Effect of Enzyme concentration

| *Test Tube No.* | *Observation* | *Time* |
|---|---|---|
| 1 | | |
| 2 | | |
| 3 | | |

## 3. Effect of Temperature

| *Test Tube No.* | *Observation* | *Conclusion* |
|---|---|---|
| 1 | | |
| 2 | | |
| 3 | | |

**4. Effect of pH**

| *Test Tube No.* | *Observation* | *Conclusion* |
| --- | --- | --- |
| 1 | | |
| 2 | | |
| 3 | | |

**5. Effect of an Inhibitor**

| *Observation* | *Conclusion* |
| --- | --- |
| | |

## Post-Lab Questions

1. Both carbon monoxide and oxygen bind at the same active site in hemoglobin. However, carbon monoxide binds so strongly that oxygen cannot displace it. What kind of an inhibitor is carbon monoxide?

2. The experiment with rennin simulates the environment in the stomach. How would the digestion of milk in the stomach be influenced by concentration and pH?

3. How does the pH experiment simulate the environment in the mouth?

4. The enzyme of a bacterium is optimal at 40 °C. Would this enzyme be more active or less active at normal body temperature or when a person has a fever?

# Use of the Laboratory Gas Burner

A ready source of heat in the chemistry laboratory is through the use of a Tirrill or Bunsen burner. In general since chemical reactions proceed faster at elevated temperatures, the use of heat enables the experimenter to accomplish many experiments quicker than would be possible at room temperature. The laboratory burner illustrated in Fig. A1.1 is typical of the burners used in most general chemistry laboratories.

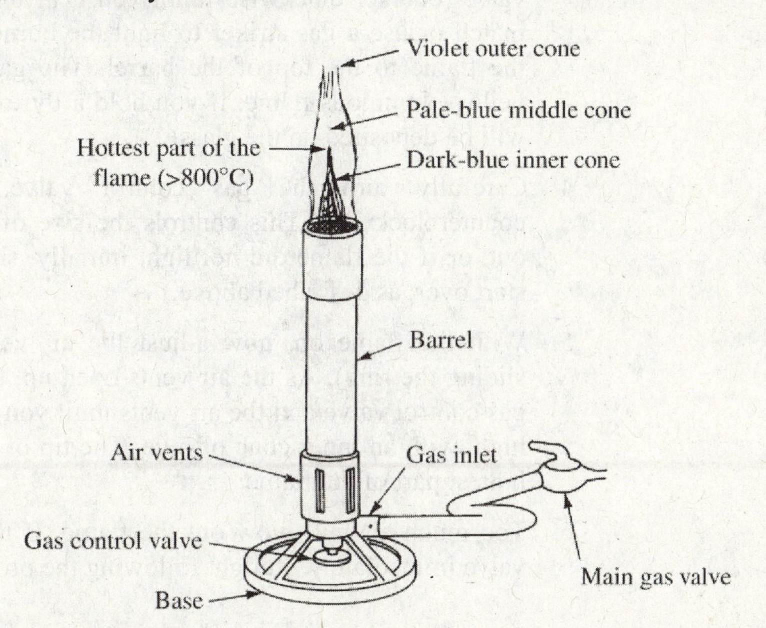


**Figure A1.1**
*The Tirrill (Bunsen) Burner.*

The burner is designed so as to allow gas and air to mix in a controlled manner. The gas often used is "natural gas," mostly the highly flammable and odorless hydrocarbon methane ($CH_4$). When ignited, the flame's temperature can be adjusted by altering the various proportions of gas and air. The gas flow can be controlled either at the main gas valve or at the needle valve at the base of the burner. Manipulation of the air vents at the bottom of the barrel allows air to enter and mix with the gas. The hottest flame is the violet outer cone. The flame also has a pale-blue middle cone and a dark-blue inner cone; the air vents, in this case, are opened sufficiently to assure complete combustion of the gas. Lack of air produces a "cooler" yellow luminous flame. This flame lacks the inner cone and most likely is smoky, often depositing soot on objects it contacts. Too much air blows out the flame.

You can see how the Bunsen burner works if you follow the directions below.

1. Before connecting the Bunsen burner to the gas source, examine the burner and compare it to Fig. A1.1. Be sure to locate the gas control valve and the air flow adjuster and see how they work.

2. Connect the gas inlet of your burner to the main gas valve by means of a short piece of thin-walled rubber tubing. Be sure the tubing is long enough to provide some slack for movement on the bench top. Close the gas control valve; if your burner has a screw-needle valve, turn the knob in a clockwise manner. Close the air vents. You can do this by rotating the barrel of the burner (or sliding the ring over the air vents if your burner is built this way).

3. Turn the main gas valve to the open position. Slowly open the gas control valve (counter clockwise) until you hear the hiss of gas. Quickly strike a match or use a gas striker to light the burner. With a lighted match, hold the flame to the top of the barrel. The gas will light. You will have a yellow, luminous flame. If you hold a Pyrex® test-tube in this flame, soot will be deposited on the glass.

4. Carefully turn the gas control valve, first clockwise and then counterclockwise. This controls the size of the flame. (If the flame goes out, or if the flame did not light initially, shut off the main gas valve and start over, as described above.)

5. With the flame on, now adjust the air vents by rotating the barrel (or sliding the ring). As the air vents open up, the flame gets blue. Adjust the gas control valve and the air vents until you obtain a flame about 3 or 4 in. high, with an inner cone of blue. The tip of the pale blue inner cone is the hottest part of the flame.

6. Too much air will blow out the flame. If this occurs, close the main gas valve immediately. Relight following the procedure in step 3.

**Figure A1.2**
*The flame rises away from the burner.*

7. Too much gas pressure will cause the flame to rise away from the burner and "roar" (Fig. A1.2). If this happens, reduce the gas flow by closing the gas control valve until a proper flame results.

8. "Flash back" sometimes occurs. Here the burner has a flame at the bottom of the barrel. Quickly close the main gas valve. Allow the barrel to cool. Relight following the procedures in step 3.

# Use of the Spectroline® Pipet Filler

**CAUTION!**

**Never use your mouth when pipetting any liquid.**

1. Pipetting any liquid with your mouth is extremely dangerous. Many liquid chemicals and solutions are poisonous or can cause burns. That is why it is important to learn to use pipetting aids such as the Spectroline® pipet filler.
2. Examine the Spectroline® pipet filler and locate the valves marked "A," "S," and "E" (Fig. A2.1). These operate by pressing the flat surfaces between the thumb and forefinger.

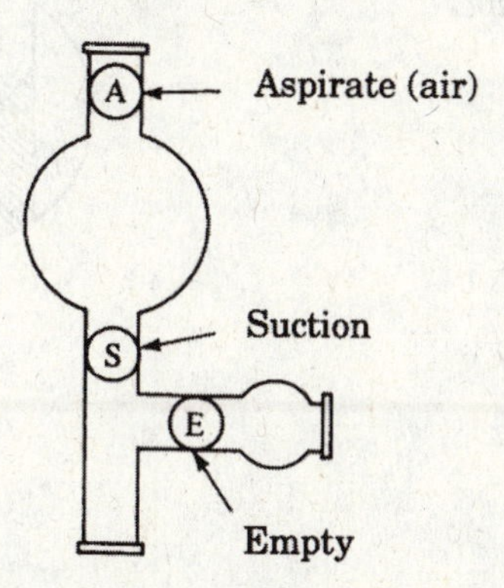


**Figure A2.1**
*The Spectroline® pipet filler.*

3. Squeeze the bulb with one hand while you press valve "A" with two fingers of the other hand. The bulb flattens as air is expelled. If you release your fingers that are pressing valve "A" when the bulb is flattened, the bulb remains collapsed.
4. Carefully insert the pipet end into the Spectroline® pipet filler (Fig. A2.2). The end should insert easily and not be forced.

**CAUTION!**

**Before inserting the pipet end into the Spectroline® pipet filler, lubricate the glass by rubbing the opening with a drop of water or glycerine.**

5. Place the tip of the pipet into the liquid to be pipetted. Make sure that the tip is below the surface of the liquid at all times
6. With your thumb and forefinger, press valve "S." Liquid will be drawn up into the pipet. By varying the pressure applied by your fingers, the rise of

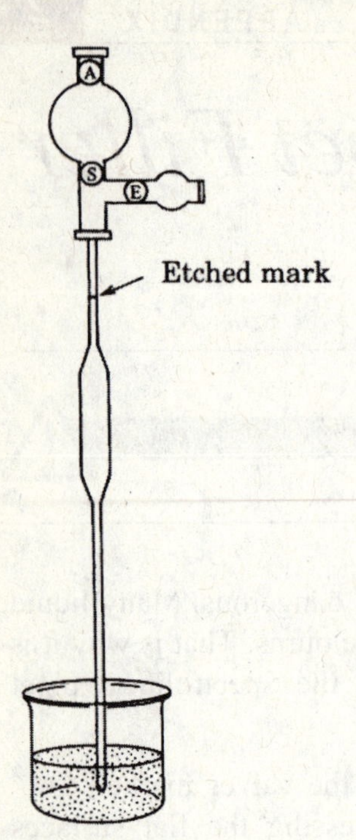


**Figure A2.2**
*Using the Spectroline® pipet filler to pipet.*

the liquid into the pipet can be controlled. Allow the liquid to fill the pipet to a level slightly above the etched mark on the stem. Release the valve; the liquid should remain in the pipet.

7. Withdraw the pipet from the liquid. Draw the tip of the pipet lightly along the wall of the beaker to remove excess liquid.
8. Adjust the level of the meniscus of the liquid by carefully pressing valve"E." The level should lower until the curved meniscus touches the etched mark (Fig. A2.3). Carefully draw the tip of the pipet lightly along the wall of the beaker to remove excess liquid.
9. Drain the liquid from the pipet into a collection flask by pressing valve "E." Remove any drops on the tip by touching the tip of the pipet against the inside wall of the collection flask. Liquid should remain inside the tip; the pipet is calibrated with this liquid in the tip. Do not blow out the excess.

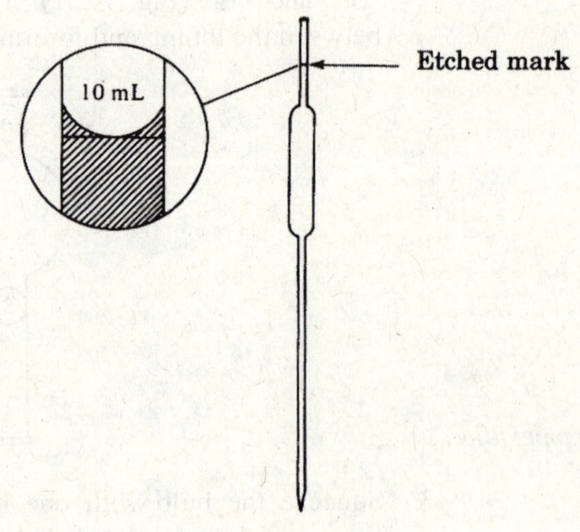


**Figure A2.3**
Adjusting the curved meniscus of the liquid *to the etched mark.*

# List of Apparatus and Equipment in Student's Locker

## AMOUNT AND DESCRIPTION

(1) Beaker, 50-mL
(1) Beaker, 150-mL
(1) Beaker, 250-mL
(1) Beaker, 400-mL
(1) Beaker, 600-mL
(1) Bunsen or Tirrill burner
(3) Clamp, extension and holder
(1) Ring clamp
(1) Cylinder, graduated by 0.1 mL, 10-mL
(1) Cylinder, graduated by 1 mL, 50-mL
(6) Dropper, medicine with rubber bulb
(2) Flask, Erlenmeyer, 50-mL
(2) Flask, Erlenmeyer, 125-mL
(2) Flask, Erlenmeyer, 250-mL
(1) File, triangular
(1) Forceps
(1) Funnel, Büchner, small
(1) Funnel, short stem
(1) Gauze, wire
(2) Glass rod
(1) Litmus paper, blue
(1) Litmus paper, red
(1) Meniscus reader
(1) Ruler
(1) Spatula, stainless steel, micro
(1) Spatula, stainless steel, regular
(1) Sponge
(1) Spot plate, white with 12 depressions
(1) Striker or box of matches

(12) Test tube, 13 × 100 mm
(12) Test tube, 16 × 125 mm
(1) Test tube brush, small
(1) Test tube brush, large
(1) Test tube holder
(1) Test tube rack
(1) Thermometer, 150 °C
(1) Tongs, crucible
(1) Wash bottle, plastic
(2) Watch glass, 6 in. and 3 in.

# *List of Common Equipment and Materials in the Laboratory*

Each laboratory should be equipped with hoods and safety-related items such as fire extinguisher, fire blankets, safety shower, and eye wash fountain. The equipment and materials listed here for 25 students should be present in each laboratory.

Acid tray
Aspirators (splashgun type) on sink faucet
Balances, triple beam (or Centigram®) or top-loading
Barometer
Clamps, extension
Clamps, thermometer
Clamps, utility
Containers for solid chemical waste disposal
Containers for liquid organic waste disposal
Corks
Detergent for washing glassware
Drying Oven
Filter paper
Glass rods, 4 and 6 mm OD
Glass tubing, 6 and 8 mm OD
Glycerol (glycerine) in dropper bottles
Hot plates
Ice maker
Paper towel dispensers
Pasteur pipets
Rings, support, iron, 76 mm OD
Ring stands
Rubber tubing, pressure
Rubber tubing, latex (0.25 in. OD)
Water, deionized or distilled
Weighing dishes, polystyrene, disposable, 73 × 73 × 25 mm
Weighing paper

# Special Equipment and Chemicals Preparations for Each Experiment

In the instructions below, every time a solution is to be made up in "water," you must use *distilled water*.

## 1 EXPERIMENT 1

*Laboratory Measurements*

### SPECIAL EQUIPMENT

| | |
|---|---|
| (25) | 50-mL graduated beakers |
| (25) | 50-mL graduated Erlenmeyer flasks |
| (25) | 100-mL graduated cylinders |
| (25) | Metersticks or rulers, with both English and metric scale |
| (25) | Hot plates |
| (25) | Single pan, triple beam balances (Centogram®) |
| (5) | Platform triple beam balances |
| (2) | Top loading balances |

## 2 EXPERIMENT 2

*Density Determination*

### SPECIAL EQUIPMENT

| | |
|---|---|
| (4) | 250-mL beakers (labeled for unknown solids) |
| (25) | Solid wood blocks, rectangular or cubic |
| (25) | Spectroline® pipet fillers |
| (26) | Pipet Pumps |
| (25) | 10-mL volumetric pipets |

### CHEMICALS

| | | |
|---|---|---|
| **Unknown Solids** | (100 g) | Aluminum, pellets or rod |
| | (100 g) | Glass beads |
| | (100 g) | Iron, pellets or cut nails |
| | (100 g) | Lead, shot |
| | (100 g) | Tin, pellets or strips |

| | | |
|---|---|---|
| | (100 g) | Zinc, pellets |
| **Unknown Liquids** | (250 mL) | Ethanol |
| | (250 mL) | Ethylene glycol (commercial permanent anti-freeze may be used instead) |
| | (250 mL) | Hexane |
| | (250 mL) | Milk, homogenized, regular |
| | (250 mL) | Olive oil |
| | (250 mL) | Sea water: this can be a 3.5% NaCl solution by weight; dissolve 8.75 g of NaCl in enough water to make 250 mL |

3 **EXPERIMENT 3**

## *Classes of Chemical Reactions*

### SPECIAL EQUIPMENT

| | |
|---|---|
| (1 box) | Wood splints |

### CHEMICALS

| | |
|---|---|
| (25 pieces) | Aluminum foil (2 × 0.5 in. each) |
| (50 pieces) | Aluminum wire (1 cm each) |
| (25 pieces) | Copper foil (2 × 0.5 in. each) |
| (25) | Pre 1982 copper penny (optional) |
| (20 g) | Ammonium carbonate, $(NH_4)_2CO_3$ |
| (20 g) | Potassium iodide, KI |
| (40 g) | Potassium iodate, $KIO_3$ |
| (20 g) | Calcium turnings, Ca |
| (20 g) | Magnesium ribbon, Mg |
| (20 g) | Mossy zinc, Zn |
| (20 g) | Lead shot, Pb |

**All the following solutions should be placed in dropper bottles.**

| | |
|---|---|
| (100 mL) | 3 M hydrochloric acid, 3 M HCl: 25 mL concentrated HCl (12 M HCl) diluted with enough ice cold water to make 100 mL |
| (100 mL) | 6 M hydrochloric acid, 6 M HCl: 50 mL concentrated HCl (12 M HCl) diluted with enough ice cold water to make 100 mL |
| (100 mL) | 3 M sulfuric acid, 3 M $H_2SO_4$: 16.7 mL concentrated $H_2SO_4$ (18 M $H_2SO_4$) is slowly added to 60 mL ice cold water; stir slowly and dilute with enough ice cold water to make 100 mL. |

(100 mL) 3 M sodium hydroxide, 3 M NaOH: dissolve 12 g of NaOH in enough water to make 100 mL solution

**In preparing the above solutions, rubber gloves, a rubber apron, and face shield should be worn. Do all preparations in the hood.**

(100 mL) 0.1 M silver nitrate, 0.1 M $AgNO_3$: dissolve 1.70 g $AgNO_3$ in enough water to make 100 mL solution

(100 mL) 0.1 M sodium chloride, 0.1 M NaCl: dissolve 0.58 g NaCl in enough water to make 100 mL solution

(100 mL) 0.1 M sodium nitrate, 0.1 M $NaNO_3$: dissolve 0.85 g $NaNO_3$ in enough water to make 100 mL solution

(100 mL) 0.1 M sodium carbonate, 0.1 M $Na_2CO_3$: dissolve 1.24 g $Na_2CO_3 \cdot H_2O$ in enough water to make 100 mL solution

(100 mL) 0.1 M potassium nitrate, 0.1 M $KNO_3$: dissolve 1.01 g $KNO_3$ in enough water to make 100 mL solution

(100 mL) 0.1 M potassium chromate, 0.1 M $K_2CrO_4$: dissolve 1.94 g $K_2CrO_4$ in enough water to make 100 mL solution

(100 mL) 0.1 M barium chloride, 0.1 M $BaCl_2$: dissolve 2.08 g $BaCl_2$ in enough water to make 100 mL solution

(100 mL) 0.1 M copper(II) nitrate, 0.1 M $Cu(NO_3)_2$: dissolve 1.88 g $Cu(NO_3)_2$ in enough water to make 100 mL solution

(100 mL) 0.1 M copper(II) chloride, 0.1 M $CuCl_2$: dissolve 1.70 g $CuCl_2 \cdot 2H_2O$ in enough water to make 100 mL solution

(100 mL) 0.1 M lead(II) nitrate, 0.1 M $Pb(NO_3)_2$: dissolve 3.31 g $Pb(NO_3)_2$ in enough water to make 100 mL solution

(100 mL) 0.1 M iron(III) nitrate, 0.1 M $Fe(NO_3)_3$: dissolve 4.04 g $Fe(NO_3)_3 \cdot 9H_2O$ in enough water to make 100 mL solution

## 4 EXPERIMENT 4

## *Physical Properties of Chemicals: Melting Point, Sublimation, and Boiling Point*

### SPECIAL EQUIPMENT

(1 roll) Aluminum foil

(1 bottle) Boiling chips

(1) Commercial melting point apparatus (if available)

(25) Glass tubing, 20 cm segments

(25) Hot plates

(100) Melting point capillary tubes

(50) Rubber rings (cut 0.25-in. rubber tubing into narrow segments)

(25) Thermometer clamps

(25) Thiele tube melting point apparatus

## CHEMICALS

(20 g) Acetamide
(20 g) Acetanilide
(20 g) Adipic acid
(20 g) Benzophenone
(20 g) Benzoic acid
(20 g) *p*-Dichlorobenzene
(20 g) Naphthalene, pure
(50 g) Naphthalene, impure: mix 47.5 g (95%) naphthalene and 2.5 g (5%) charcoal powder
(20 g) Stearic acid

**The following liquids should be placed in dropper bottles.**

(200 mL) Acetone
(200 mL) Cyclohexane
(200 mL) Ethyl acetate
(200 mL) Hexane
(200 mL) Methanol (methyl alcohol)

## 5 EXPERIMENT 5

## *Factors Affecting Rate of Reactions*

## SPECIAL EQUIPMENT

(5) Mortars
(5) Pestles
(25) 10-mL graduated pipets
(25) 5-mL volumetric pipets

## CHEMICALS

**Solutions should be put into dropper bottles. In preparing the solutions, wear a face shield, rubber gloves, and a rubber apron. Do in the hood.**

(500 mL) 6 M HCl: add 258 mL concentrated HCl (12 M HCl) to 200 mL water. Mix and bring it to a volume of 500 mL.

(100 mL) 2 M $H_3PO_4$: add 13.3 mL concentrated $H_3PO_4$ (15 M $H_3PO_4$) to 50 mL ice cold water. Mix and bring it to a volume of 100 mL.

(100 mL) 6 M $HNO_3$: add 50.0 mL concentrated $HNO_3$ (12 M $HNO_3$) to 50 mL ice cold water. Mix and bring it to a volume of 100 mL.

(100 mL) 6 M acetic acid: add 34.4 mL glacial acetic acid (99-100%) to 50 mL water. Mix and bring it to a volume of 100 mL.

| | |
|---|---|
| (100 mL) | 3 M $H_2SO_4$: add 16.7 mL concentrated $H_2SO_4$ (18 M $H_2SO_4$) to 60 mL ice cold water. Mix slowly and bring it to a volume of 100 mL. |
| (100 mL) | 6 M $CH_3COOH$: add 35.3 mL glacial acetic acid to 40 mL ice cold water. Mix slowly and bring it to a volume of 100 mL. |
| (500 mL) | 0.1 M $KIO_3$: **Caution!** *This solution must be fresh. Prepare it on the day of the experiment.* Dissolve 10.7 g $KIO_3$ in enough water to make 500 mL volume. |
| (250 mL) | 4% starch indicator: add 10 g soluble starch to 50 mL cold water. Stir it to make a paste. Bring 200 mL water to boil in a 500-mL beaker. Pour the starch paste into the boiling water. Stir and cool to room temperature. |
| (500 mL) | 0.01 M $NaHSO_3$: dissolve 0.52 g $NaHSO_3$ in 100 mL water. Add slowly 2 mL concentrated sulfuric acid. Stir and bring it to a volume of 500 mL. |
| (250 mL) | 3% $H_2O_2$, 3% hydrogen peroxide: use a commercial bottle that is available from any pharmacy |
| (150) | Mg ribbons, 1 cm long |
| (25) | Zn ribbons, 1 cm long |
| (25) | Cu ribbons, 1 cm long |
| (25 g) | Manganese dioxide, $MnO_2$ |
| (25 g) | Marble chips |

## 6 EXPERIMENT 6

# *The Law of Chemical Equilibrium and the Le Chatelier Principle*

## SPECIAL EQUIPMENT

| | |
|---|---|
| (2 rolls) | Litmus paper, blue |
| (2 rolls) | Litmus paper, red |

## CHEMICALS

**The following chemicals should be provided in dropper bottles.**

| | |
|---|---|
| (50 mL) | 0.1 M copper(II) sulfate, 0.1 M $CuSO_4$: dissolve 0.80 g $CuSO_4$ (or 1.25 g $CuSO_4 \cdot 5H_2O$) in enough water to make 50 mL |
| (50 mL) | 1 M ammonia, 1 M $NH_3$: dilute 3.3 mL concentrated $NH_3$ (28%) with enough water to make 50 mL. **In the preparation wear a face shield, rubber gloves, and a rubber apron. Do in the hood.** |
| (25 mL) | Concentrated HCl, 12 M HCl |
| (100 mL) | 1 M hydrochloric acid, 1 M HCl: add 8.5 mL concentrated HCl (12 M HCl) to 50 mL ice water; add enough water to make 100 mL. **In the preparation wear a face shield, rubber gloves and a rubber apron. Do in the hood.** |

| | |
|---|---|
| (150 mL) | 0.1 M phosphate buffer: dissolve 1.74 g $K_2HPO_4$ in enough water to make 100 mL. Dissolve 1.36 g $KH_2PO_4$ in enough water to make 100 mL. Mix 100 mL of the $K_2HPO_4$ solution with 50 mL of $KH_2PO_4$ solution. |
| (100 mL) | 0.1 M potassium thiocyanate, 0.1 M KSCN: dissolve 0.97 g KSCN in enough water to make 100 mL |
| (100 mL) | 0.1 M iron (III) chloride, 0.1 M $FeCl_3$: dissolve 2.7 g $FeCl_3 \cdot 6H_2O$ (or 1.6 g $FeCl_3$) in enough water to make 100 mL |
| (100 mL) | Saturated saline solution: add 290 g NaCl to warm (60°C) water. Stir until dissolved. Cool to room temperature. |
| (50 mL) | 1.0 M cobalt chloride, 1.0 M $CoCl_2$: dissolve 11.9 g $CoCl_2 \cdot 6H_2O$ in enough water to make 50 mL |

7

## EXPERIMENT 7

# *pH and Buffer Solutions*

## SPECIAL EQUIPMENT

| | |
|---|---|
| (5) | pH meters |
| (12 rolls) | pHydrion paper® (pH range 0 to 12) |
| (5 boxes) | Kimwipes® |
| (5) | Wash bottles |
| (100) | 10-mL graduate pipets |
| (25) | Spot plates |
| (25) | 10-mL beakers |

## CHEMICALS

| | |
|---|---|
| (250 mL) | 0.1 M acetic acid, 0.1 M $CH_3COOH$: dissolve 1.4 mL glacial acetic acid in enough water to make 250 mL |
| (500 mL) | 0.1 M sodium acetate, 0.1 M $CH_3COONa$: dissolve 6.8 g $CH_3COONa \cdot 3H_2O$ in enough water to make 500 mL |
| (1 L) | 0.1 M hydrochloric acid, 0.1 M HCl: add 8.3 mL concentrated HCl (12 M HCl) slowly to 100 mL ice water, with stirring; dilute with enough water to 1 L. **Prepare in the hood; wear a face shield, rubber gloves, and a rubber apron.** |
| (1 L) | 0.1 M sodium bicarbonate, 0.1 M $NaHCO_3$: dissolve 8.2 g $NaHCO_3$ in enough water to make 1 L |
| (500 mL) | Saturated carbonic acid, $H_2CO_3$: use a bottle of club soda or seltzer water; these solutions are approximately 0.1 M carbonic acid. |

**The following solutions should be placed in dropper bottles.**

(100 mL) 0.1 M HCl, prepared above

(100 mL) 0.1 M ammonia, 0.1 M $NH_3$: dilute 0.7 mL concentrated $NH_3$ (28%) with enough water to make 100 mL

(100 mL) 0.1 M sodium hydroxide, 0.1 M NaOH: dissolve 0.4 g NaOH in enough water to make 100 mL solution

## 8 EXPERIMENT 8

## *Structure in Organic Compounds: Use of Molecular Models*

### SPECIAL EQUIPMENT

Color of spheres may vary depending on the set; substitute as necessary.

**A. Alkanes.**

(50) Black spheres - 4 holes
(150) White (or Yellow) spheres - 1 hole
(50) Colored spheres (Green) - 1 hole
(25) Red (or Blue) spheres - 2 holes
(200) Sticks
(25) Protractors
(75) Flexible grey bonds (or Springs)

**B. Cyclohexane.**

(175) Black spheres - 4 holes
(350) White (or Yellow) spheres - 1 hole
(25) Colored spheres (Green) - 1 hole
(525) Sticks

## 9 EXPERIMENT 9

## *Aspirin: Preparation and Properties (Acetylsalicylic Acid)*

### SPECIAL EQUIPMENT

(25) Büchner funnels (65 mm OD)
(25) Filtervac or no. 2 neoprene adapters
(1 box) Filter paper (5.5 cm, Whatman no. 2)
(25) 250-mL filter flasks
(25) Hot plates

## CHEMICALS

| | |
|---|---|
| (1 jar) | Boiling chips |
| (25) | Commercial aspirin tablets |
| (100 mL) | Concentrated sulfuric acid (18 M $H_2SO_4$) (in a dropper bottle) |
| (100 mL) | 1% iron(III) chloride: dissolve 1 g $FeCl_3 \cdot 6H_2O$ in enough distilled water to make 100 mL (place in a dropper bottle) |
| (100 mL) | Acetic anhydride (use a freshly opened bottle) |
| (300 mL) | Ethyl acetate |
| (100 g) | Salicylic acid |

# 10 EXPERIMENT 10

## *Carbohydrates*

## SPECIAL EQUIPMENT

| | |
|---|---|
| (50) | Medicine droppers |
| (125) | White spot plates with 25 depressions |
| (2 rolls) | Litmus paper, red |

## CHEMICALS

| | |
|---|---|
| (1 bottle) | Boiling chips |
| (400 mL) | Fehling's reagent (solutions A and B, from Fisher Scientific Co.) |
| (200 mL) | 3 M sodium hydroxide, 3 M NaOH: dissolve 24.00 g NaOH in 100 mL water and then add enough water to bring to 200 mL. |
| (200 mL) | 2% starch solution: place 4 g soluble starch in a beaker. With vigorous stirring, add 10 mL water to form a thin paste. Boil 190 mL water in another beaker. Add the starch paste to the boiling water and stir until the solution becomes clear. Store in a dropper bottle. |
| (200 mL) | 2% sucrose: dissolve 4 g sucrose in enough water to bring to 200 mL. Store in a dropper bottle. |
| (50 mL) | 3 M sulfuric acid, 3 M $H_2SO_4$: add 8.5 mL concentrated sulfuric acid (18 M $H_2SO_4$) to 30 mL ice cold water; **pour the sulfuric acid slowly along the walls of the beaker, this way it will settle on the bottom without much mixing**; stir slowly in order not to generate too much heat; when fully mixed, bring the volume to 50 mL. **Wear a face shield, rubber gloves, and a rubber apron when preparing. Do in the hood.** Store in a dropper bottle. |
| (100 mL) | 2% fructose: dissolve 2 g fructose in enough water to make 100 mL. Store in a dropper bottle. |

| | |
|---|---|
| (100 mL) | 2% glucose: dissolve 2 g glucose in enough water to make 100 mL. Store in a dropper bottle. |
| (100 mL) | 2% lactose: dissolve 2 g lactose in enough water to make 100 mL. Store in a dropper bottle. |
| (100 mL) | 0.01 M iodine in KI: dissolve 1.2 g potassium iodide, KI, in 80 mL water. Add 0.25 g iodine, $I_2$. Stir until the iodine dissolves. Add enough water to make 100 mL. Store in a dark dropper bottle. |

## 11 EXPERIMENT 11

## *Fats and Oils: Preparation and Properties of a Soap*

### SPECIAL EQUIPMENT

| | |
|---|---|
| (25) | Büchner funnels (85 mm OD) |
| (25) | No. 7 one-hole rubber stoppers |
| (1 box) | Filter paper (Whatman no. 2) 7.0 cm |
| (1 roll) | pHydrion® paper (pH range 0 to12) |
| (25) | Pasteur pipettes |

### CHEMICALS

| | |
|---|---|
| (1 bottle) | Boiling chips |
| (100 mL) | 1% $Br_2$ solution in cyclohexane: mix 1 mL $Br_2$ with 99 mL of cyclohexane. *Prepare fresh solutions prior to each laboratory period; distribute in 4 dropper bottles; do not store.* **Wear face shield, rubber gloves, and rubber apron during preparation. Prepare in the hood.** |
| (¼ lb) | Butter, unsalted |
| (100 mL) | 5% calcium chloride, 5% $CaCl_2$: dissolve 5 g $CaCl_2 \cdot H_2O$ in enough water to make 100 mL. Store in a dropper bottle. |
| (1 L) | Corn oil |
| (150 mL) | Dichloromethane, $CH_2Cl_2$ |
| (1 L) | 95% ethanol, $CH_3CH_2OH$ |
| (100 mL) | 5% magnesium chloride, 5% $MgCl_2$: dissolve 5 g $MgCl_2$ in enough water to make 100 mL. Store in a dropper bottle. |
| (100 mL) | Mineral oil. Store in a dropper bottle. |
| (1 L) | Saturated sodium chloride, NaCl: dissolve 360 g NaCl in enough water to make 1 L |
| (1 L) | 25% sodium hydroxide, 25% NaOH: dissolve 250 g NaOH in enough water to make 1 L |

12 **EXPERIMENT 12**

# *Separation of Amino Acids by Paper Chromatography*

## SPECIAL EQUIPMENT

| | |
|---|---|
| (1) | Drying oven, 105–110 °C |
| (2) | Heat lamps or hair dryers |
| (25) | Whatman no. 1 chromatography paper, 15 × 8 cm |
| (25) | Rulers, metric scale |
| (25) | Polyethylene, surgical gloves |
| (150) | Capillary tubes, open on both ends |
| (1 roll) | Aluminum foil |
| (2) | Wide-mouth jars |

## CHEMICALS

| | |
|---|---|
| (25 mL) | 0.12% aspartic acid solution: dissolve 30 mg aspartic acid in 25 mL distilled water |
| (25 mL) | 0.12% phenylalanine solution: dissolve 30 mg phenylalanine in 25 mL distilled water |
| (25 mL) | 0.12% leucine solution: dissolve 30 mg leucine in 25 mL distilled water |
| (25 mL) | Aspartame solution: dissolve 150 mg Equal® sweetener powder in 25 mL distilled water |
| (50 mL) | 6 M hydrochloric acid, 6 M HCl, solution: place 10 mL ice cold distilled water into a 50-mL volumetric flask. Add slowly 25.0 mL of concentrated HCl (12 M HCl) and bring it to volume with distilled water. **Wear a face shield, rubber gloves, and a rubber apron when preparing. Do in the hood.** |
| (1 L) | Solvent mixture: mix 600 mL 1-butanol with 150 mL acetic acid and 250 mL distilled water. **Solvent mixture *must* be freshly prepared the day it is to be used.** |
| (1 can) | Ninhydrin spray reagent (0.2% ninhydrin in ethanol or acetone); do not use any reagent older than 6 months old. |
| (1 can) | Diet Coca-Cola® |
| (4 packets) | Equal® (or NutraSweet®) sweetener |
| (10 g) | Iodine crystals, $I_2$ |

## 13 EXPERIMENT 13

# *Isolation and Identification of Casein*

## SPECIAL EQUIPMENT

| | |
|---|---|
| (25) | Hot plates |
| (25) | 600-mL beakers |
| (25) | Büchner funnels (O. D. 85 mm) in no.7 1-hole rubber stopper |
| (7 boxes) | Whatman no. 2 filter paper, 7 cm |
| (25) | Rubber bands |
| (25) | Cheese cloths (6 × 6 in.) |

## CHEMICALS

| | |
|---|---|
| (1 jar) | Boiling chips |
| (1 L) | 95% ethanol, $CH_3CH_2OH$ |
| (1 L) | Diethyl ether:ethanol mixture (1:1) |
| (0.5 gal.) | Milk, homogenized, regular |
| (500 mL) | Glacial acetic acid |

**The following solutions should be placed in dropper bottles:**

| | |
|---|---|
| (100 mL) | Concentrated nitric acid (12 M $HNO_3$) |
| (100 mL) | 2% albumin suspension: dissolve 2 g albumin in enough water to make 100 mL |
| (100 mL) | 2% gelatin: dissolve 2 g gelatin in enough water to make 100 mL |
| (100 mL) | 2% glycine: dissolve 2 g glycine in enough water to make 100 mL |
| (100 mL) | 5% copper(II) sulfate: dissolve 5 g $CuSO_4$ (or 7.85 g $CuSO_4 \cdot 5H_2O$) in enough water to make 100 mL |
| (100 mL) | 5% lead (II) nitrate: dissolve 5 g $Pb(NO_3)_2$ in enough water to make 100 mL |
| (100 mL) | 5% mercury(II) nitrate: dissolve 5 g $Hg(NO_3)_2$ in enough water to make 100 mL |
| (100 mL) | Ninhydrin reagent: dissolve 3 g ninhydrin in enough acetone to make 100 mL. Do not use a reagent older than 6 months. |
| (100 mL) | 10% sodium hydroxide: dissolve 10 g NaOH in enough water to make 100 mL |
| (100 mL) | 5% sodium nitrate: dissolve 5 g $NaNO_3$ in enough water to make 100 mL |
| (100 mL) | 1% tyrosine: dissolve 1 g tyrosine in enough water to make 100 mL |

14 EXPERIMENT 14

# Enzymes

## SPECIAL EQUIPMENT

| | |
|---|---|
| (25) | Hot plates |
| (100) | Medicine droppers |
| (10) | Stop watches |
| (25) | White spot plates, 25 depressions |

## CHEMICALS

| | |
|---|---|
| (20 g) | Liver, raw, cut into small, pea-size chunks |
| (20 g) | Liver, cooked, cut into small, pea-sized chunks |
| (500 mL) | Milk, homogenized, regular |

**The following solutions should be placed into dropper bottles.**

| | |
|---|---|
| (300 mL) | 1% α-amylase solution: use α-amylase from barley malt with a specific activity of 100,000 units (available from Sigma-Aldrich® Chemicals). Dissolve 3 g of α-amylase in enough water to make 300 mL. *Prepare fresh the day of the experiment.* |
| (200 mL) | Buffer, pH 2 (available commercially as a standardized pH buffer solution) |
| (200 mL) | Buffer, pH 4 (available commercially as a standardized pH buffer solution) |
| (200 mL) | Buffer, pH 7 (available commercially as a standardized pH buffer solution) |
| (200 mL) | Buffer, pH 9 (available commercially as a standardized pH buffer solution) |
| (200 mL) | Buffer, pH 12 (available commercially as a standardized pH buffer solution) |
| (100 mL) | 3 M HCl, 3 M hydrochloric acid: 25 mL concentrated HCl (12 M HCl) diluted with enough ice cold water to make 100 mL |
| (200 mL) | 3% $H_2O_2$, 3% hydrogen peroxide (available from any pharmacy) |
| (200 mL) | 2% rennin solution: use Junket® Rennet tablets. This is available in the pudding section of the supermarket. It may be ordered directly from Redco Foods, Inc., P.O. Box 1027, one Hansen Island, NY 13365 (1-800-556-6674). Five tablets are approx. 4 g; pulverize to a powder with a mortar and pestle. Add to 50 mL of water and stir to dissolve (the solution will be cloudy). Add enough water to make 200 mL. *Prepare fresh the day of the experiment.* |

(200 mL) 1% starch solution: place 2 g soluble starch in a beaker. With vigorous stirring, add 10 mL water to form a thin paste. Boil 190 mL water in another beaker. Add the starch paste to the boiling water and stir until the solution becomes clear.

(200 mL) 2% starch solution: place 4g soluble starch in a beaker. With vigorous stirring, add 10 mL water to form thin paste. Boil 190 mL water in another beaker. Add the starch paste to the boiling water and stir until the solution becomes clear.

(100 mL) 0.01 M iodine in KI: dissolve 1.2 g potassium iodide, KI, in 80 mL water. Add 0.25 g iodine, $I_2$. Stir until the iodine dissolves. Add enough water to make 100 mL. Store in a dark dropper bottle.

(100 mL) 0.1 M $Pb(NO_3)_2$, 0.1 M lead(II) nitrate: dissolve 3.31 g $Pb(NO_3)_2$ in enough water to make 100 mL